NORTH CAROLINA
STATE BOARD OF COMMUNITY COLLEGES
LIBRARIES
ASHEVILLE-BUNCOMBE TECHNICAL COLLEGE

DISCARDED

APR 15 2025

Fighting HAZARDOUS MATERIAL Fires

A Selection of Recent Articles from **FIRE ENGINEERING** "Leading the fire service since 1877"

Fire Engineering is published monthly by
Technical Publishing, a company of The
Dun & Bradstreet Corporation.

875 Third Avenue, New York, N.Y. 10022

Copyright © 1983 by Technical Publishing Co.,
a company of The Dun & Bradstreet Corp.
All rights reserved. No part of this publication may be reproduced or
transmitted in any form without permission in writing from the publisher.

Printed in the United States of America

Library of Congress Catalog Number: 83-50529
ISBN 0-912212-00-4

Contents

Number of Trucks Moving Hazardous Cargoes Increases Odds for Accident	1
Four-Alarm Fire at Chemical Plant Causes Decontamination Problems	4
Be Prepared for Decontamination at Hazardous Materials Incidents	6
Hazardous Cargoes Make Rail Yards Prime Objects for Emergency Plans	9
Railroad Tanker Improves Center's Haz-Mat Training	12
Unique Hazard Posed by Oxygen-Enriched Atmosphere	13
Illinois Department Prepared for Haz-Mat Emergencies	15
Contaminants Released at Fire Require 27 Days for Cleanup	16
Ammonia Leak	19
Houston F.D. Haz-Mat Team Finds Activity Rises as Reputation Spreads	21
Entire Buildings are Labeled with NFPA 704M Marking System	24
Railroad Goes to Fire Departments with Hazardous Materials Program	26
Dump Fire Complicated by Generation of Chlorine Gas, Hydrochloric Acid	30
240,000 Forced to Flee Chlorine Released in Canadian Train Wreck	32
Toxic Spill From Tank Car Causes Evacuation of Mile-Square Area	34
Computer Put in Vehicle of Haz-Mat Unit	37
On-Site Gasohol Blending Breeds Widespread Extinguishing Problems	38
Toxic Flammable Chemicals, Gases Breed Trouble in Electronic Plants	41
New Inspection Program Tracks Hazardous Materials	43
The Process of Making Decisions at Hazardous Materials Incidents	45
Add an EMS Specialist to your Haz-Mat Team	50
Delayed Alarm at Chemical Fire	53
Unstable Chemical Controlled Safely	57
Plan Works, Leak Stopped in Liquid Nitrogen Tank	59

Preface

Incidents involving hazardous materials make up perhaps the biggest challenge facing fire departments today. The variety of ways hazardous materials can spill or leak or burn or react is staggering. But more threatening to fire department responders is the enormous quantity of these substances being transported by road, rail, plane and pipeline.

This widespread transportation of hazardous materials places extraordinary risk at the doorstep of every community, no matter how small. Any other type of emergency incident is affected by community size. But not a hazmat incident. Anyone can test their community's exposure to dangerous substances by spending an hour or so watching the trucks and trains passing through and by noting the warning placards on them. Those vehicles may not even be scheduled to stop in the community, but sometimes they do...unexpectedly...tragically.

The main defense a community and a fire department has is awareness and training. That's why these articles from Fire Engineering have been collected here. They explore a subject worth a second look. Each one has something unique and important to offer. It is hoped that the reader never will confront a major incident involving hazardous materials as described here. But the odds are not so optimistic...

Jerry W. Laughlin, Editor
Fire Engineering

Photo by Warren Isman

Compressed-gas cylinders from an overturned truck. Contents of different cylinders could leak and mix. Officers should establish specific objectives before beginning to clear sites.

Number of Trucks Moving Hazardous Cargoes Increases Odds for Accident

BY CHIEF WARREN E. ISMAN
Director
Department of Fire and Rescue Services
Montgomery County, Md.

The area protected by every fire department has a potential for having a hazardous materials incident involving truck transportation. No matter how small the community, the movement of hazardous goods by truck occurs frequently.

For those who still do not think that an incident is possible, consider the frequency of transportation of gasoline to the local service station, bulk propane to the storage facility and propane cylinders to the homes and businesses of the community. Fertilizers and pesticides can be brought into any community but they are particularly abundant in rural areas. Additionally, there is the local food store which has goods such as chlorine bleach, aerosol cans and insecticides, all brought in by truck.

The size of the problem

In order to indicate the magnitude of the problem, the following basic statistics are provided:
- Over four billion tons of hazardous materials are shipped (including reshipment) each year.
- The majority of these shipments are petroleum products. There are approximately 100,000 cargo tank shipments a day of gasoline alone. Parked end to end, these tanks would stretch from Washington, D.C., to Chicago.
- There are approximately 100,000 shippers of hazardous materials and 80,000 carriers (mainly trucks) transporting hazardous materials of all types.
- It is estimated that there are 354,000 generators of hazardous waste, which will increase the truck transportation problem significantly.
- There are over 180 million shipments of hazardous materials annually.
- There are 175,000 portable 2000-pound tanks which move by truck, 150,000 tank trucks, 80,000 tank semi-trailers, 50,000 large portable tanks, and an unknown number of vans moving hazardous materials.

DOT regulation

Interstate transportation of hazardous materials by truck is regulated by the United States Department of

Transportation (DOT). Many states have adopted these same regulations for intrastate transportation as well. However, a 1979 special study of the National Transportation Safety Board (NTSB) determined six basic reasons why those involved in hazardous materials shipments do not always comply with federal regulations:

1. The complexity of the regulations themselves.
2. The complexity of the industry interrelationships.
3. Economic pressures.
4. Lack of awareness of the regulations by industry personnel.
5. Lack of personnel training.
6. Indifference.

On the basis of this study, NTSB recommended improved regulation of cargo tank integrity, control of liquid surge in tank-truck transportation and regulation of routing to reduce the risks of transportation accidents.

NTSB has also reported that: "The number of accidents involving hazardous materials in transportation each year are not known. From 1971 through 1979, 95,167 incidents were reported to the DOT, about 90 percent of them by motor carriers.

"The adequacy and relevancy of much of the data in the incident reports are questionable," and "the credibility of the available incident data is questionable and there is no routine validation of the data (by the DOT)."

Some not placarded

DOT estimates that between 5 and 15 percent of all trucks on the road at any time carry hazardous materials. A survey by the Virginia Department of Transportation Safety found that 65 percent of the materials transported were flammable or combustible liquids, about 10 percent of the trucks were carrying more than 1000 pounds of hazardous materials and thus required placarding, 41 percent of the trucks requiring placarding were either not placarded or incorrectly placarded, and 23 percent of the trucks carrying hazardous materials failed to carry the required shipping papers.

It is certain that no matter what the size of your department, you must be prepared for handling a hazardous materials incident involving a truck. Unfortunately, many senior fire officers at this point throw up their hands and say they have insufficient funds to get the necessary equipment. Other actions are more useful.

Preparing for a hazardous materials incident involving a truck requires planning. This will involve an investment of time by senior fire service personnel, but there is very little direct cost. Next, some basic control materials as well as a resource document need to be prepared. Finally, the remainder of the department needs to be trained.

Gasoline tanker collided with a cement truck. Leaking fuel was diked and covered with foam.

The plan

The plan for handling a truck incident should consist of:

1. A map of the road networks, starting with the major roads and working downward to the less traveled ones. The plans should include a drawing of the area covered, sewer systems, topography and drainage, water supply availability along the road (pressure and static sources) and exposure problems.

2. A survey of the trucking companies, commodities shipped and routes used to transport hazardous materials. This information will form the basis for developing the reference and resource list. It can be prepared by actually observing the trucks that come through the community, noting the name of the carrier and contacting them for further information. In addition, the users of hazardous materials in the community can be surveyed to determine the chemicals used, the carrier which delivered them and the frequency of shipment.

3. A reference manual which contains the phone numbers of key emergency personnel; local, state and federal organizations capable of providing assistance; and private organizations, businesses and individuals that may provide help.

4. Aids for decision-making at the incident, including maps of utilities, sewers and transportation routes; topographical maps; diagrams of cargo tanks; copies of the DOT placards and labels in current use; and mathematical conversion tables.

5. A disaster plan organizational program. The disaster plan should show an organizational structure for various types of truck incidents; a chart showing the command structure, particularly using officers and personnel from mutual aid departments; description of who has legal authority for putting the plan into effect; and an outline of other local organizations and their responsibilities at an incident.

Control equipment

Equipment for controlling spills and leaks does not have to be elaborate. Much of the equipment can be developed by the fire fighters themselves.

For example, small holes in tank trucks can be stopped or reduced by using tapered wooden plugs of various sizes. In addition, rubber sheets can be cut into gaskets, with a hole made in the middle for placement on a large butterfly bolt. The bolt is then inserted in the hole along with the rubber gasket and tightened against the inside of the tank.

Photo by Lt. Leonard King, Montgomery County Dept. of Fire & Rescue Services.
Fuel oil truck in a precarious position and leaking from a break in its side. Spill was diked.

Photo by Montgomery County, Md. Police Department
Improper welding without complete purging of this tank resulted in an explosion. Windows were broken in a 1/2-mile area.

Photo by Lt. Leonard King, Montgomery County Dept. of Fire & Rescue Services
Sulfuric acid container fell from truck and broke open. Soda ash dike was formed.

In addition, some common tools can be carried to help control leaks. These would include:
Pipe wrenches (various sizes)
Spark-resistant wrenches and hammer
Open end and box wrenches
Crescent wrenches
Bung wrenches
Wire cutters
Pliers
Screwdrivers (flat and Phillips)
Hand drill and bits
Wood auger and bits
Chisels (wood and metal)
Hack saw and blades
Packing gland wrench
Tire snips
Banding tool and steel band
Tubeless tire patching kit
Rubber sheeting
Hardwood plugs
Rags and paper towels
Nonhardening gasket compound
Epoxy glue
Valve sealant
Packing (Teflon)
Tape (Teflon, electrical, plastic)
Silicone sealant
Lead wool
"O" rings
Sheet metal screws
Nuts, bolts, washers, and lockwashers (3/8, 1/2, 5/8, 3/4")
Pipe fitings:
Pipe nipples
Elbows
Unions
Reducers
Couplers
Caps
Plugs
Valves
Administrative:
Binoculars
Paper (regular and graph)
Rulers
Pencils, pens, markers
Tags,
Chemical flares
Flashlight and spare batteries
Rubber bands
Sample containers (various sizes and construction)

Putting this equipment together will not involve a great cost. However, no mention has been made yet of regular and special protective equipment. Remember, full protective gear including positive-pressure self-contained breathing apparatus is necessary. It is also important to point out that many chemicals require additional special protective equipment. Fully enclosed acid or chemical suits, natural rubber gloves and boots, or radiological monitors may be needed. If your department does not have this specialized equipment, then your resource book should indicate where it can be located. If specialized equipment is called for and you do not have it available, evacuate civilian and emergency personnel until help can be obtained.

There are times when specialized extinguishing agents will be necessary. Again, if you do not have sufficient foam, alcohol-type foam or class D agents, then locate them and record the information in your resource document.

Check claims

Several commercial companies have developed leak-stopping, plugging, diking and neutralizing equipment. Advertisements for these products have begun to appear in fire service magazines. If you are interested, find out from the manufacturer the fire departments which have used the product. Then check to determine if the results are really as good as the manufacturer claims.

Obviously, no fire department can be prepared for an incident, based upon your plan, in a classroom setting. The plan, responsibilities and tactics can be discussed in a group. While this method takes little preparation time, it is usually only good as a review.

Another technique is to give the students a description of a possible truck incident in the locality. Using their knowledge of the plan, the students must develop the positioning of apparatus, use of personnel, notifications for other agencies, use of reference material and anticipation of further problems.

The use of models is another excellent way to develop coordinating skills. Railroad models of buildings or homemade models can be used to represent a portion of the local community. Using a street layout with model trucks and cars (painted cotton to simulate fire, smoke and spills), a problem can be developed. The students then must determine the tactics and strategy of handling the incident.

Simulation

Finally, training can take the form of a disaster drill involving a simulated incident. A truck can be parked with a wrecked vehicle placed nearby. Injuries can be simulated using makeup. The vapor cloud can be simulated using either smoke bombs or carbon dioxide discharged from cylinders. Local, state and federal government agencies can assist as they would at an actual incident. Private agencies such as the Red Cross or local radio operators can be asked to participate as they would at an actual incident. This type of simulation can be as basic or elaborate as desired.

The transportation of hazardous materials by truck is increasing. The probability of an incident occurring is also increasing. Fire departments, therefore, need to be prepared.

Preparation for a truck incident involves developing a plan, a strategy, some specialized equipment and a training program.

The problem won't go away. You should begin your preparation before the incident, and not try to do it while you are in the middle of a major problem. ❏ ❏

Fire apparatus along U.S. Route 1 (bottom) drafted from the canal when hydrant supplies proved inadequate. Only one narrow road, just behind elevating platform, led to plant.

Four-Alarm Fire at Chemical Plant Causes Decontamination Problems

BY ROBERT BURNS
Staff Correspondent

A four-alarm fire at a chemical plant in Lawrence Township, N.J., included a series of explosions that ripped the roof off one building and sent a column of black smoke so high that it was visible for 10 miles. Several hundred nearby residents were evacuated to avoid potentially toxic fumes.

During the attack by fire fighters, disputes arose with the New Jersey Department of Environmental Protection and the New Jersey State Police over who was in charge of the scene.

Lawrence Township, located just north and bordering the state capital at Trenton, occupies 20 square miles and has a population of 32,000. The eastern edge of the township along Alternate U.S. 1 is a heavy commercial area. A bit farther east, along the Delaware and Raritan Canal, are a pair of chemical plants dealing in hazardous materials.

Fire protection is furnished by three fire companies: Slackwood Fire Company, Lawrence Road Fire Company and the Lawrenceville Fire Company. Each unit operates independently in its area under its own chief. Each company has three pumpers. Slackwood is equipped with an 85-foot elevating platform and Lawrenceville with a 100-foot ladder.

The fire broke out at the Saturn Chemical Company, located at 1600 New York Ave. in an industrial section adjacent to Route 1, at 2:32 a.m. last Aug. 21.

Familiar location

After the alarm was transmitted, Dale Robbins, second assistant chief of first-due Slackwood company, saw the towering column of black smoke while he was en route to the station. He correctly surmised that the fire was at the Saturn Chemical Company. The plant was no stranger to the fire fighters. They have been to three major blazes at this location in the past seven years, which prompted Chief Rudy Fuessel of Slackwood to prepare an extensive mutual-aid plan.

Robbins immediately ordered both the Lawrence Road and Lawrenceville companies to respond. Upon receipt of the alarm via radio, according to the plan, neighboring Hamilton Township automatically dispatched one pumper.

Saturn Chemical produces resins and thinners for industrial and commercial paints. Within the plant were 27 tanks for the storage of mineral spirits, xylene, styrene, vinyl toluene and butyl acrylate—all ingredients for a potential disaster. Adjoining the plant to the south was the Hydrocarbon Research, Inc. plant with two 125-foot-high cracking units. Hundreds of drums of flammable hydrocarbons were stored in the yard directly adjacent to the fire.

Access and water problems

The location posed problems of access and water supply for arriving fire fighters. Only one narrow road led directly to the fire. Slackwood's first-due engine went directly to the Saturn Chemical yard and hooked up to a hydrant, only to find there was not enough water to supply it. Robbins ordered the first-due Lawrence Road pumpers to lay in with 4 and 3-inch lines from a hydrant approximately 1000 feet from the fire. Slackwood's second pumper was to hook up to a street hydrant. Lawrenceville, commanded by Chief Earl Wilbur, also laid in with a 4-inch line approximately 1500 feet to a 10-inch water main on New York Ave.

Robbins directed his aerial platform

into the yard to operate directly on the fire and had Prospect Heights' ladder pipe placed in service to cool the barrels in the yard of the Hydrocarbon Company. Wilbur also positioned Lawrenceville's 100-foot ladder in the rear of the yard to protect exposed tanks.

Water supply officer

Assistant Chief James Yates of Lawrence Road set up two pumpers along New York Ave. and had a deluge gun from one pumper play on the involved tanks at Saturn. Yates also was designated by Robbins to be the water supply officer.

When Hamilton Fire Company arrived on U.S. 1 to draft from the canal directly beside the road, Chief Don Kanka called for Hamilton's ladder in order to place a ladder pipe in operation. Kanka then called for mutual aid from the Enterprise Fire Company, Colonial Fire Company and Rusling Hose's ladder, as well as additional assistance from the Hamilton Fire Company.

Slackwood requested West Trenton's aerial platform and Penning Road's hose wagon with 2000 feet of 4-inch hose. Orders were to draft from the canal and operate ladder pipes across the area onto the exposed tanks. Responding units also stretched 4-inch lines across 6-inch girders spanning the 35-foot-wide, 6-foot-deep body of water. Deluge guns were carried across the narrow span of girders and a line was laid across to supply Lawrenceville's ladder.

This operation resulted in a complete shutdown of U.S. 1 in both directions.

Fire fighters faced a fire in 12 8000-gallon-capacity tanks that had been in a corrugated metal building. The building was destroyed in a blast before the arrival of any of the fire companies. Robbins' main concern in addition to controlling the fire was the protection of 13 tanks of solvents just 30 feet from the main fire, and a 75×200-foot shed 50 feet to the north containing hundreds of 55-gallon drums of product.

Outside interference

The New Jersey Division of Environmental Protection (DEF) under Paul Giradina, director of health management, responded from nearby Trenton. DEP hampered the operation by refusing to allow the owners of the plant into the fire area. This was of serious concern to Robbins and Fuessel because the owners were the only people in a position to advise the fire fighters as to what hazards were involved.

The DEP personnel also ordered fire fighters to set up deluge guns and then get "out of the area," an order which the chiefs refused to obey.

Samples of air taken by DEP showed the presence of hydrogen cyanide, a dangerous poison, and xylene and styrene, both suspected carcinogens. However, Giradina was able to determine that all of these were at less than half of the levels considered dangerous.

Evacuation

Lawrence Township Civil Defense Director John Kubilewiz ordered the evacuation of that portion of the township east of Alternate U.S. 1 as a precaution against potentially deadly fumes and the possibility of a major explosion. Police evacuated about eight blocks of residents who were directed to the Slackwood Fire station as an evacuation center. They were allowed to return to their homes about 5 p.m.

Companies approaching from the west could not reach the canal to draft, and hydrants in the immediate area had sparse supplies. Water Liaison Officer Yates, in attempting to solve the water problem, directed Prospect Heights to lay 900 feet each of 3 and 2½-inch line in relay with the DeCou Fire Company, which laid a 4-inch line 1000 feet to a hydrant located on a 16-inch main on the west side of Alternate U.S. 1.

This resulted in the complete shutdown of both U.S. 1 and Alternate U.S. 1, and created a major traffic problem in the area for the New Jersey State Police, Lawrence Township and Trenton Police, Mercer County Sheriff's Department and the Mercer County Fire Police.

Police issue orders

At one point, the New Jersey State Police "ordered" Fuessel to shut down the operation on U.S. 1 to relieve the traffic jam, another order with which the chief refused to comply. When Fuessel declared the fire under control at 4:35 p.m., U.S. 1 was reopened to traffic.

Mercer County Airport dispatched its foam unit to the fire but it was not placed in service because its approach was blocked by early arriving pumpers. Hamilton Township's hazardous material response team was on location and stood by in proximity suits, but their services were not required.

Numerous mutual-aid companies covered vacated stations in the area, with the City of Trenton sending a pumper to the Slackwood station.

Slackwood remained on the scene until 11:35 p.m. Later Slackwood discovered it had lost 1000 feet of 4-inch hose, 1200 feet of 3-inch and 600 feet of 1¾-inch hose which could not be reclaimed, as well as 25 pairs of boots and nine Nomex coats that had to be destroyed. They had a major clean-up job to do on their apparatus, as did other responding companies.

Three fire fighters were treated for smoke inhalation. They were the only injuries reported.

Plant owners estimated their loss at $2 million. The owner stated the plant would be rebuilt on its present location. DEP representatives emphasized that more stringent construction and chemical handling standards would be required before a permit for a new plant would be issued.

The cause of the fire is still under investigation. ▫ ▫

Be Prepared for Decontamination At Hazardous Materials Incidents

BY JOHN R. LEAHY, JR.
*District Chief
Largo, Fla., Fire Department*
and
ROGER A. McGARY
*Chief
Takoma Park, Md., Volunteer Fire Department*

Fire officers should always have a concern for the proper decontamination of personnel who have been involved at hazardous materials incidents. Decontamination is that activity where the individual, clothing, apparatus and equipment are placed in a safe condition, based on the type of materials involved in the incident.

In each response and exposure, the officer in charge must identify the product involved, being alert to the nature of the exposure (leak, spill, fire) and the nature of the product (liquid, solid or gas). Different conditions clearly require different tactics.

As an officer in charge develops the tactics for a given incident, he must be alert to the potentials of smoke and water contamination, as well as the products of decomposition that may be in the immediate area of the incident. He must also review whether the material is toxic, corrosive, radioactive, etc.

Based on the judgment made, the most effective tactic may well be to avoid exposure, letting the material burn while removing all personnel and equipment to a safe area. Officers should recognize that many toxic materials involved in a fire may be rendered neutral by the extreme heat developed. On the other hand, an attack on a given incident may result in a cooling action which produces a heavier contamination within the smoke as well as the water runoff. A more extensive site cleanup may be required when an attack is made.

Major factor: weather

After the incident has been brought to a conclusion there is then the concern for the decontamination of personnel and equipment involved. A major deciding factor in the specific procedure used is the weather. Excessive heat or excessive cold will affect a decision to contaminate either at the scene or a remote location.

It is highly recommended that the officer in charge assign an officer to supervise decontamination. Hopefully, this activity will be coordinated through your emergency medical system. Industrial hygienists or medical consultants may be desired. A pre-incident plan for decontamination should be developed, with a list of resources required for that activity.

The officer in charge and the individual assigned to supervise decontamination must identify the contaminant and how it reacts with the body, whether by inhalation, ingestion or skin absorption. This is where the medical consultant or industrial hygienist can play a critical role. Your hazardous materials resource directory should be of assistance in this situation.

Select proper method

Once identification has been accomplished, the next logical step would be to determine the proper method of decontamination.

Will it be washing a liquid from the individual and his protective clothing with water? Or will it be a vacuum cleaner for dust or soot removal? We should avoid using a hose line on protective clothing that is covered with solid materials. The spray may scatter the solid materials, resulting in inhalation by anyone near by.

Remove breathing apparatus, protective clothing and even work clothing when no water is available. The key here is that the contaminant may penetrate any clothing, with the material being absorbed through the skin or causing a chemical burn to the individual.

Contaminants flushed off with water or vacuumed off must then be contained for proper disposal.

Of major concern in any decontamination effort is how personnel will be transferred to a decontamination or medical facility. It is our recommendation that a single vehicle be used if at all possible. This avoids multiple vehicle contamination. In the pre-incident plan, you should determine the availability of a bus for this activity. You may decide to have it dispatched immediately for specific incidents. If a bus is not readily available, is there a large rescue vehicle in which a number of individuals could be moved from the incident to the decontamination center?

Below 50 degrees

We recommend that the following concepts be considered if the temperature is below 50 degrees:
● Transport to a decontamination center with protective clothing on, or
● If the contaminant is of such a serious nature that there is a respiratory problem, then it may be advisable to remove the protective clothing prior to transport (the clothing may well hold

Photos by Asst. Chief Michael May, Hyattstown, Md., Vol. F.D.
Wet by sprinklers, fire fighters are restricted to isolation area while decisions on possible exposure to hazardous materials and need for decontaminations are made.

Fire fighters doff environmental suits while awaiting determination of decontamination needs.

the respiratory contaminant and release it within the confined space of the mode of transportation), or

• Transport personnel with their breathing apparatus on. Concerns for distance of transport and adequate air supply must be addressed if this mode is chosen.

With each of these options, the officer in charge must be alert to the time required to obtain the transport vehicle, whether the personnel are wet and experiencing chill, and whether the material is wet or dry.

Removing clothing with dry materials on them at the scene, prior to transport, is advisable. Care must be taken to avoid creating a dust condition. Personnel should keep breathing apparatus on during the clothing removal. Clothing and breathing apparatus must be bagged and held for later inspection and decontamination—or destruction.

Those personnel assisting should be protected, too. This may be done with a filter-type respirator, gloves and throwaway garments.

At the scene

When decontaminating at the scene, regardless of temperature, a specific area should be established for this activity. It should be identified, barricaded and restricted to entry. Following decontamination of personnel, that area must also be decontaminated.

The method of transportation (with or without protective clothing and breathing apparatus) is directly related to the type of contaminant and the probability of inhalation of that contaminant or skin absorption by the contaminant penetrating the clothing. In extremely cold temperatures personnel may suffer shock from any hosing and removal of clothing done at the scene—unless immediate warm transportation is available.

What would you use for your decontamination center? The following are recommended: a hospital, fire station, training center, industrial locations, swimming pool with use of the wading area, a school. Regardless of the site selected, the officer in charge must be concerned with the collection of any runoff water used for decontamination and the decontamination of that facility.

If you choose a hospital, be alert to the potential expense involved. A recent survey of Montgomery County, Md., hospitals resulted in a quote by one of $40 to $100 per person, dependent on what would be required. Others would not even offer an estimate because most are not prepared to handle decontamination.

If the ambient temperature is above 50 degrees, we would recommend that initial decontamination occur at the scene, followed by transportation to a decontamination facility. Initial decontamination would result in the individual being hosed down, with the appropriate removal and collection of protective clothing, respiratory protection and work clothing. The individual would then be transported to the previously selected decontamination center.

Techniques

The overall technique for decontamination may vary, but we recommend these steps for consideration:

1. Hose or shower with water from above and down over the individual, never at a 90-degree angle.
2. Remove clothing in this order: the helmet, disconnect the low-pressure hose and remove the air pack, gloves, coat (keeping hands off the exterior shell; a helper with gloves may be required), boots and bunker pants, the facepiece, work clothing and underwear.
3. Soap and shower twice.
4. Body check by a qualified technician.
5. Replace clothing. When an individual has been decontaminated, you should make provisions for clothing. This could be in the form of the paper throwaway type or coveralls. It is highly recommended that all fire fighters keep a complete set of clothing in their own locker for such incidents.
6. Begin treatment.
7. Health monitoring.

Neutralization is required where acid or alkalies are involved. The treatment may be necessary at the scene on an immediate basis when the individual

Staging area where fire fighters remain with their unit is a key control of incident.

An **incident commander** gives direction to fire fighters dressed in environmental suits prior to their approach to a chemical spill.

comes in contact with the material. Then a follow-up would be necessary at the decontamination center. Acids or alkalies can penetrate the normal fire fighter protective clothing, resulting in burns to the body. In a situation involving acids and alkalies, modesty cannot be considered. The diluted but not neutralized corrosive material can lie and collect in the waistband of underwear or in socks and shoes of the individual. Many have sustained second and third-degree burns from corrosives for failing to immediately remove their clothing.

Medication and definitive treatment are as prescribed by the medical adviser. We cannot overemphasize the need for proper identification of the material so that the medical personnel can prescribe the proper medication. An incident that occurred last Feb. 27 demonstrated this. A chemical spill at the Environmental Laboratories in Hanover, Va., a northern Richmond suburb, sent 24 people to the hospital. The material was first identified as diborane, a flammable gas used in the making of rocket fuel. Later tests indicated that the material was pentaborane, another rocket fuel ingredient "10 times more lethal than diborane."

Health monitoring

To effectively monitor personnel within your department, you first must develop a data base from a pre-employment or membership physical that includes the appropriate blood gas analysis.

When an incident has occurred, and following decontamination, your medical adviser should conduct a thorough examination of the individual to establish the current data. Your medical adviser then has a reference point and an exposure point. Significant changes in these two would give an indication of toxicological effect.

This current, or post-incident, data would include noting in an appropriate record the date, time, material, its physical properties and the time of exposure, as well as appropriate blood gas work, blood pressure, respirations and pulse, urine analysis and dosimeter check.

Depending on the material involved, the medical adviser may want to make an additional medical evaluation of the same items a week, month or even six months from the given incident.

We cannot overemphasize the need to maintain adequate records of any exposure. Each fire fighter should have an appropriate log within his personnel jacket that would indicate each hazardous materials exposure incident. It may be the option to maintain this record with the medical record. Some jurisdictions have separated such records for confidentiality. This is a point where the medical consultant must be involved.

Equipment decontamination

Meanwhile, back at the site. All clothing, if contaminated, should be bagged for containment and confinement until appropriate analysis can be completed. This may well require a gas chromatographic analysis of the clothing based on the contaminant.

SCBA should be treated in the same way, i.e., bagged, contained and confined. The situation may require replacement of diaphragms, straps and facepieces.

Tools, appliances and hose are a major problem. They should be secured in a proper facility while further investigation is done. Tools and other metal appliances can undoubtedly be hosed down. Decontamination of hose will be a much more difficult problem. The fabric itself can be impregnated with the contaminant, holding it for a long period of time.

When we consider decontamination of apparatus, it does provide a mind-boggling situation. There are many nooks and crannies where contaminant particles may lie. They can be in the seats, the hose bed, the pump panel, etc. We would recommend that the unit be appropriately hosed down and confined until experienced specialists can be sought to check for any hidden contaminants.

The fire scene itself requires expertise to evaluate the problem. We have a responsibility to the people and environment. There should be concern for water removal, the water table itself, the saturated ground and the debris.

Financial responsibility

Whenever we attempt to have such cleanup done, the officer in charge must recognize the financial responsibilities involved. Simply calling in a commercial hazardous materials recovery team without securing proper financial clearance could result in a major expense to the department or jurisdiction involved.

As you consider the various supplies required, you may want to review the materials used to decontaminate. For the most part, water is the number one item, but many materials are not water soluble and another procedure must be used. If water is used, then how do you hold, contain, neutralize and remove it from the scene or a decontamination center?

Use of a cut-off 55-gallon drum, a child's wading pool, a portable drafting tank or specially designed holding tanks may be required. Hospitals and industrial locations in your area may already have diversion and holding tanks for just this purpose. Chemical plants usually have chemical sewer systems that collect all chemical spills into a central holding basin for neutralization.

If you have a solid material and have used a vacuum cleaner, how do you handle the decontamination of that unit and disposal of solid material? If absorbents were used, what procedures are used to dispose of them? In both cases it may require special permits to use sanitary landfills. Or the materials may be incinerated if such facilities exist.

In the overall process of decontamination of protective and work clothing the officer in charge must consider the equipment used to wash these items. Is there a chance of residual contaminants staying in that machinery? Experts should be found to inspect such equipment after use. In some situations it may be advisable to destroy all clothing worn. This could be by landfill burial or incineration.

If technical help is not found prior to an incident, personnel may be subjected to health problems which could have been avoided if a decontamination procedure had been developed. ☐ ☐

Hazardous Cargoes Make Rail Yards Prime Objects for Emergency Plans

BY GENE P. CARLSON

Many aspects of preplanning are discussed in training and during drills in the fire service. One area that has not been touched upon is railroad switchyards, large or small, that dot our nation's rail system.

Emergency response officials are aware of the vast amounts of hazardous materials and ordinary combustibles transported by rail. How many, however, have considered the handling of these commodities while they are temporarily staying in the local railroad yard?

Past experience has indicated that serious incidents can occur. Among the more serious incidents which have occurred in railroad terminals are those in the accompanying table.

Preplanning of yards and terminals can be accomplished only by meeting with the railroad terminal officials and developing an adequate emergency plan.

Yardmaster's importance

At a rail yard, the key individual is the yardmaster. He is in charge of the railroad switching operations and will normally be the initial emergency coordinator until relieved of duties by a higher railroad official. The yardmaster usually has experience as a switchman and is well acquainted with the yard. He also has access to the yard communications network, directs all switching movements, and controls trains entering and leaving the terminal. He is a key individual in the planning due to his knowledge, access to information and material, and the fact that he will initiate the emergency plan if an incident occurs.

In some cases, a railroad terminal will be made up of more than one yard. These may be adjacent to each other, either side by side or end to end, or in some cases, separated by several miles in a metropolitan area.

In the case of a large terminal with more than one yard, the hierarchy of management would proceed up from the yardmaster to a general yardmaster and/or terminal manager, who supervises the yardmasters. There is then a terminal trainmaster, who is responsible for the terminal operating personnel and their activities, and finally the terminal superintendent, who is responsible for the overall operation of the terminal.

Place	Date	Estimated Loss in Millions	Deaths	Product
Decatur, Ill.	7/19/74	$18	7	Isobutane
East St. Louis, Ill.	1/22/72	$7.5	—	Propylene
Houston, Tex. (Englewood)	9/21/74	$13	1	Butadiene
Roseville, Calif.	4/28/73		—	Bombs
Wenatchee, Wash.	8/6/74	$7.5	2	Monomethyl-amine Nitrate

Numerous problems in fire and spill control exist in classification and switchyards.

The degree to which these men are involved in the planning will depend on local circumstances, but in many cases they will be the railroad authority to endorse the effort, coordinate railroad personnel's involvement, approve the final plan, and issue the order for its adoption and use.

Selection of routes

The preplanning must start with establishing fire department response routing to and into the yard. Maps or aerial photographs should be obtained which clearly denote roadways in the yard and illustrate all key locations. This should indicate access routes, evacuation lanes, roadways, major obstructions, meeting points, pertinent structures, possible command post and staging area locations, yardmaster's office, water supplies, drainage systems, isolation tracks, fuel and hazardous material storage areas, and cutoff controls. All information, of course, will have to be kept current with construction and other changes in the yard.

An essential point in response planning is the selection of several locations to which apparatus should respond and meet railroad personnel who will direct them to the incident. Several meeting points may be necessary because of the size of a yard, the possibility of incidents at different locations, the need to reach both sides of an incident, or the need to provide for alternate access routes to the yard. Fire service personnel should recommend additional fire lanes or roads, if necessary, to obtain adequate access to high-potential areas.

Fire department dispatchers may have to direct companies to specific meeting points as circumstances vary with the incident. Companies should respond to assigned meeting points unless directed elsewhere. Railroad employees should be assigned to each meeting area.

Consider obstructions

It may be necessary to develop alternative response routes because of blockage of grade crossings or yard roadways by long trains. Detail any obstructions to response or operations, such as fences, ditches, power lines and elevated or below-grade trackage. Be sure to cover both the front and back of the yard. This may require the response of more than one fire department.

Don't overlook main line tracks which may have higher speed, through-trains moving on them, fire hydrants, or the dual usage of routes for both response and escape. If possible, escape routes should be different from response routes. When laying out response routes within a yard, a description of the basic yard operations should be obtained.

An early consideration during an emergency will be the establishment of a command post and identification of the products in the cars involved. Identification of the products can often be aided by locating the command post in the yardmaster's office. This office is ideal because it has the capability of identifying rail cars, their contents, and their locations in the yard. It also has waybills and possibly a computer terminal to obtain product identification. Remember that in a yard, switch crews may not have the papers to assist in lading identification.

Good view of yard

An additional advantage of using the yardmaster's office for the command post in many terminals is that it is elevated and provides an excellent vantage point for viewing much of the yard. Also, the yardmaster's office has direct access to the interyard communication system, including the yard speaker system, contact with switch engines, the switchmen's walkie-talkies, and other key railroad personnel.

The possibility of a command post in other than the yardmaster's office must be included in the planning. In all cases, the command post should not be too close or too remote from the incident, will need plenty of working room, and adequate communications. The preplan must be sure to include procedures for alerting all participants of the location of the command post.

Communications are vital in a rail yard. The incident commander and command post must be able to work closely with railroad personnel. It will be necessary to determine what cars are on adjacent tracks, to have locomotives drop the cars they are moving on a remote track and stand by to clear tracks adjacent to the emergency or to remove hazardous materials. Movement of cars may be necessary for access or safety.

In the event of an incident, an announcement should be made over both channels of the railroad communication system that units working the incident should switch to a single channel and hold the channel open for emergency operations. Communications are also necessary to control the movement of trains into and through the yard. The communication problem enlarges when the yard serves more than one railroad.

Photo by C. J. Wright

Special railroad equipment should be included in pre-fire planning list of resources.

Emergency numbers

Communications preplanning includes having the fire department acquire the emergency phone numbers that will be needed during an incident in a yard. The fire department also should know how to contact railroad officials at home after business hours. Likewise, railroad personnel should learn how to contact emergency response forces and technical assistance and provide them with helpful information that the railroad dispatcher has available.

It must be stressed that railroad personnel should record the time the incident occurred and when flame impingement on tank cars began, as well as the time the fire department was notified and when apparatus arrived. These times should be reported to the incident commander since they can be critical in making decisions involving hazardous materials. The railroad should determine who will notify the fire department and when and how the public will be informed or warned. Because of the importance of the decision to evacuate the surrounding area and the way these people will be told to leave, the determination of who will be involved in the decision-making must be detailed.

A staging area of adequate size with lighting should be selected for additional apparatus and supplies. The area should not only be accessible to units moving into the rail yard, but it also should have direct routes to most of the yard.

The yard should be surveyed for equipment that will be useful during the incident. This will include any fire protection within the yard. Hydrants may be widely spaced or totally lacking. Since in many cases only small mains exist, fire flow tests should be run.

The possibility of filling any available tank cars and using them as a water supply source in the yard should be checked. The procedure for filling and withdrawing water from these cars must be established. Bottom outlet cars will have to be used and proper adapters to fire department threads must be available. Any alternative water sources should be noted.

Through coordination with terminal officials, an isolation track, or tracks, should be designated for the safe deposition of problem cars. Consider a remote area that is also accessible by roadway. Also evaluate the exposures, including drainage, and the availability of water.

Small equipment, such as nonsparking tools that may be required to stop a leak, should be located. The availability of pickup trucks, vehicles equipped to run on rails, and switch engines to assist in moving hose lines, equipment and personnel will be important. Flat cars, gondolas, cranes or derricks may also be necessary and must be included in the list of resources. Also, the availability of operators and how they are contacted for an emergency must be outlined in the planning.

Railroad personnel

The mechanical department of the railroad should be considered as a resource. They have car men and diesel mechanics available. These personnel can be helpful in removing ignition sources throughout the yard, which could include mechanical refrigerator cars, fusees, switch lamps, and possible alcohol heater cars, or any other mechanical equipment that could add to the problems.

All possible problems should be foreseen, such as storage areas within the yard containing fuel, fusees, torpedoes, or other materials which could become involved or exposed. Shops where paint is stored or used and acetylene and oxygen cylinders should be noted. The problem of keying a radio microphone for a transmission igniting a vapor leak cannot be overlooked, along with electrical and fuel supply cutoffs

Tankers without placards indicate the difficulty of identifying products in railroad yards without any advance planning. These tankers contain lubricating oil.

for the yard.

The movement of locomotives may also have to be restricted, including those on main lines passing through the yard, to eliminate possible ignition of vapors. Knowledge of prevailing winds may be important to vapor cloud dispersion and placement of emergency response apparatus. Closing of valves or the use of non-sparking tools may have to be accomplished under the cover of water fog since there could be a buildup of static electricity that could jump to a grounded object. Fire service personnel should obtain basic familiarization with locomotives and tank cars and their valving arrangements.

The planning must include not only a fire, but also a liquid or gas leak that could be flammable, corrosive, toxic, or a combination of these. Considerations must include drainage, the location of discharges, what they expose, any adjacent bodies of surface water, and how spills will be controlled or gas leaks dispersed. The role of environmental agencies cannot be overlooked. It is necessary to determine the sources of materials to dike, inhibit, neutralize, or absorb large quantity liquid spills.

Any special fire department operations must be outlined and standard operating procedures must be developed to carry them out. This may include the assignment of additional or special equipment on the first alarm, special water supply operations, the requirement for unusual hose layouts, fire fighting operations coordinated with railroad equipment, protection of adjacent exposures, tactics for contingencies, and unusual or massive evacuations.

In all cases, specific problems will have to be correlated with the community disaster plan and those agencies involved during a large-scale incident. The finalized plan should be distributed to all agencies involved and periodically reviewed for updating.

Training necessary

Once the planning has been completed, special training should be given emergency response personnel so they are familiar with the plan, the rail yard and the operations of the railroad. They should endeavor to learn the appropriate terminology so they can converse with those railroad personnel they will work with. Any special operational procedures will need to be practiced with adequate drills.

The preplanning for a rail yard can become more complicated if there are adjacent yards or more than one railroad using a yard. This will require additional options, notification, and communications. Stopping of trains and yard movements become more difficult as do access and evacuation. In some instances, a common plan may be sufficient. However, similar individual plans, or completely different ones, may be necessary for each yard or railroad.

Because of the unique considerations at a railroad yard, planning is vital. It should include local input and involvement by appropriate yard officials so that all areas are covered. Proper planning will assure a coordinated effort carried out in a cooperative, efficient manner should an incident occur.

The author would like to acknowledge the encouragement and assistance of R. G. Kuhlmann, manager, fire prevention and hazardous materials, of Burlington Northern in the preparation of this article.

Railroad Tanker Improves Center's HazMat Training

Training for rail accidents involving hazardous materials is more realistic at the Jefferson Parish Fire Training Center in Bridge City, La., now that their own tank car is in place. The 40-year-old general purpose tank car was scheduled for scrapping until Union Tank Car Company learned that the training center could use it to demonstrate problems with all kinds of hazardous cargo.

"This tank car is invaluable to us in teaching fire fighters all we know about combating railway incidents involving hazardous materials," explained George Martinsen, coordinator of the training center. "We've been actively seeking a tank car, and we're extremely pleased that Union Tank Car acted affirmatively on our request."

Getting the tank car to the 8-acre training center near New Orleans was a considerable undertaking. When Union Tank Car Company, a Trans Union Corporation affiliate, agreed in May of 1980 to donate the car, it was in Kansas. Fire center officials then prevailed on Missouri-Pacific and Santa Fe railroad personnel to deliver the car to the Bridge City area. Avondale Shipyards then trucked the tank car to the training site and provided cranes to position it on waiting rails.

The usefulness of the tank car was demonstrated during a two-day seminar last October, when it was the focal point of a 16-hour training exercise. Instructors, led by the center's assistant coordinator, Terry Hayes, and the Jefferson Parish Fire Department's hazardous materials officer, Captain Dan Civello, put more than 100 fire fighters through their paces—everything from containing anhydrous ammonia, chlorine and bromine leaks (simulated with smoke bombs) to extinguishing LPG and hydrocarbon fires.

Hayes noted that even though much progress has been made in the safe transportation of hazardous materials, "there continues to be a great need for more education and more training in ways to control the potential dangers that exist when moving these materials by truck, rail, car, sea and pipeline."

Eliminating the unknown is the most important task a fire fighter faces when coping with hazardous materials, according to Hayes. He gives prominent mention to two items in all his lectures: binoculars and track shoes. The binoculars are used at long range to size up the situation; the track shoes are for moving quickly when it's time to "get the hell out of there." Although Hayes manages to inject some humor with this approach, he is deadly serious while teaching when to go in and when to get out of hazardous materials incidents.

Five-car train planned

Martinsen has been the driving force behind the training center since its beginning in the mid 1970s. At present, Martinsen and his assistants are concentrating on the center's "railroad."

"Union Tank Car gave us momentum by donating the first car for our railroad," Martinsen said, "and now we are pursuing our goal to obtain an LPG tank car, a caustic or acid car, a box car and a caboose to complete the five-car train we need to devise a total hazardous materials program for fire fighters." ▫ ▫

More accidental oxygen-enriched atmospheres are created during transfer and transport operations than in any other aspect of oxygen use.

Unique Hazard Posed by Oxygen-Enriched Atmosphere

BY JOHN E. BOWEN

Situations in which oxygen-enriched atmospheres (OEA) are present are potentially dangerous and becoming more common.

Although oxygen is colorless, odorless and tasteless, its oxidizing properties and associated fire hazard cause fire departments to become involved. And not all OEA incidents are unusual or spectacular ones like the 1967 flash fire that killed the three American astronauts in the Apollo training exercise or the $7 million Titan missile that exploded in its silo after a liquid-oxygen leak developed during a fuel discharge operation.

Any fire fighter could become exposed to an oxygen-enriched atmosphere. Fire fighters in Los Angeles County were routinely servicing their resuscitator oxygen cylinder in the fire station last December. A full replacement cylinder was identified by a piece of tape placed over the valve. When the tape was removed this time, adhesive clung to the valve and prevented a tight seal around the gasket. Flammable adhesive came into contact with an oxidizing agent under pressure and ignited. The oxygen-enriched flame quickly burned through the steel cylinder.

Fortunately, the cylinder was small and the fire was quickly extinguished without injury. But the fire fighters had witnessed the awesome quickness with which an OEA incident occurs.

An oxygen-enriched atmosphere can occur in many other locations where oxygen is used. Normally oxygen makes up slightly less than 21 percent of the earth's atmosphere at sea level. Thus when the oxygen concentration exceeds 21 percent, or the partial pressure of oxygen exceeds 3.09 psi (14.7 psi normal atmospheric pressure × 21 percent), an oxygen-enriched atmosphere exists. Atmospheric pressure may also be expressed as 760 mm of mercury.

An oxygen concentration of 21 percent is quite adequate for most forms of life and also for combustion. However, there are situations that have been brought into common usage by modern technology for which the usual atmospheric concentration and partial pressure of oxygen are not adequate. Therefore, the OEA comes into being.

OEA accidents

The following are other examples of OEA incidents.

An inexperienced employee was told to inert an aircraft's fuel lines with nitrogen, a common practice. Instead of nitrogen, however, he carelessly connected an oxygen cylinder to the lines. The resulting explosion killed two men and destroyed the airplane.

A hospital patient died in a fire that started within the oxygen tent covering her bed. She had tried to light a cigarette in the oxygen-enriched atmosphere of the tent.

A mechanic installed a pressure gage on an oxygen cylinder. The gage, contaminated with hydraulic fluid, exploded instantly.

The severity of the OEA problem was more formally recognized in 1969 with publication of the first edition of NFPA 53M, which address OEA hazards. Other pertinent information is to be found in academic research journals and in United States military and British Royal Air Force publications. Occasional articles on specific aspects of this subject have appeared also in NFPA periodicals. The information for this article was gleaned from these sources.

Relationships altered

The combustion reaction occurs between a fuel and an oxidizer. The reaction rate, measured in terms of flame propagation, is governed partly by the concentration of fuel and oxidizer relative to each other, by the temperature of the reactants and by the ambient atmospheric pressure. This relationship is usually altered in an OEA, so a fire in a class A or class B fuel may not behave as we have learned to expect it to under usual atmospheric conditions.

So great is this effect of elevated oxygen levels and pressures that materials generally considered to be nonflammable may suddenly become highly flammable. Slow-burning materials may be consumed with seemingly explosive quickness. In short, virtually every material will burn in 100 percent oxygen, even some fire extinguishants!

The fire and explosion hazards of flammable and combustible liquids and gases depend upon factors in addition to those listed above, such as the temperature required for formation of flammable vapor mixtures and the critical fuel concentrations for flame propagation (the lower and upper explosive or flammable limits). The lower explosive limit usually varies little with oxygen concentration but the upper explosive limit is often much higher at elevated oxygen levels. That is, the flammable range widens as the oxygen level increases. Also, autoignition temperatures typically decrease at higher oxygen concentrations.

Ignition temperatures for combustible solids are usually lower in an OEA as compared to those in 21 percent oxygen. Flame propagation rates are typically much greater in an OEA also.

Fire extinguishment, like combustion, also differs in an OEA. Additional requirements are placed upon extinguishing agents and extinguishing systems because of the increased dangers.

An OEA incident can be handled properly only if you know in advance through prefire inspections where OEAs exist. Do not allow yourself to lapse into the belief that these situations are so rare as to make them of little concern.

OEAs are used daily in hospitals and other medical facilities (operating rooms, hyperbaric chambers for victims of carbon monoxide poisoning and the "bends," incubators and resuscitators). Medical uses for OEAs are not restricted to medical facilities, though, because many outpatients use oxygen therapy equipment in their homes and offices.

Industrial applications of OEAs are common also. You'll undoubtedly be able to think of many such uses, such as welding and metal-cutting operations, chemical manufacturing, etc. Incidentally, the most serious hazards in welding and metal-cutting operations are not likely to occur in major industrial plants. Rather, small repair shops and home hobby shops are apt to hold greater dangers because safety precautions may be poorly understood or even ignored.

Medical and military

An obvious OEA environment surrounds the preparation of oxygen used in medicine and industry, and its subsequent transfer and transportation.

Almost every military and commercial aircraft has an oxygen breathing system on board, a system that provides the potential for either inflight or on-ground OEA emergencies. OEAs are also used in spacecraft and on-the-ground flight simulators, in the fueling and de-fueling of rockets and in deep-sea diving vessels.

Lastly, don't forget the OEA environments that cannot be planned for, the accidental ones. These can occur when oxygen is inadvertently substituted for some other gas, for example. A large number of people confuse the terms "air" and "oxygen," and believe that these terms and the gases are synonymous. The misunderstanding has had fatal results on several occasions.

Other potential causes of unwanted OEAs include use of oxygen in poorly ventilated spaces and leakage from oxygen apparatus.

What can you, as fire fighters, do to minimize the effects of an OEA mishap to which you are summoned! How do you handle an OEA incident?

Planning a must

Basic to the successful handling of an OEA emergency, like just about every other type of incident, is planning. You must know where these OEAs exist. Only then can you recognize that you are indeed confronted with an out-of-the-ordinary situation. Get out and conduct on-site inspections.

The second step in controlling an OEA incident is in-depth training of fire fighters and also of employees and other people who are routinely in or near an OEA area prior to the emergency. And not only must they be trained, but the training must be frequently updated and maintained at a high level of proficiency. Fires in OEAs happen with explosive rapidity. Rarely will there be sufficient time for the fire department to reach the scene before serious injury occurs and extensive property damage is done. Serious burns can be inflicted upon a person wearing cotton clothing within two seconds in 100 percent oxygen!

A fire in an OEA will burn with much greater intensity than in the ambient atmosphere. Extremely high temperatures and pressures will build up much faster in an OEA, especially in a fixed-volume OEA environment such as a sealed hyperbaric chamber. The pressure increase can cause the chamber to rupture explosively, and the abnormally high temperatures will contribute to rapid fire spread. For these reasons, a fire in an OEA cannot always be extinguished in the conventional manner.

More water needed

Water is effective on fires in class A fuels under OEA conditions—if it can be applied in sufficient quantities in a very short time. A "sufficient quantity" of water under OEA conditions is much greater than that needed in 21 percent oxygen at normal atmospheric pressure.

The technique of water application in an OEA, especially in a confined area OEA, is just as important as the volume applied, because all surfaces within the closed chamber must be covered by the minimum water spray density virtually simultaneously and immediately. This minimum density, incidentally, has been reported by some British researchers to be 1.25 gpm per square foot under their experimental conditions. It remains to be seen whether this application rate will suffice under other circumstances.

Don't count on using halogenated hydrocarbon extinguishants in OEAS, either, unless you have carefully evaluated their effects beforehand. Several that are useful in normal atmospheres fail in OEAs for various reasons. Some may actually be flammable themselves in an OEA.

Bromotrifluoromethane (Halon 1301) appears to be the most promising of these extinguishants for use in OEAs. However, even Halon 1301 must be applied at much higher than usual concentrations in the OEA. And herein lies another potential problem, because Halon 1301 may be toxic at high concentrations.

No studies have been published concernting the effectiveness of carbon dioxide, dry chemicals and low-expansion foams in an OEA environment. High expansion foams show promise, but it takes too long to apply them with today's equipment. □ □

Improper tightening and contamination by particles have often been the cause of accidental oxygen-enriched atmospheres—*photos by the author.*

Apparatus and equipment of Normal's Hazardous incident response team.

Illinois Department Prepared For Haz-Mat Emergencies

BY CHIEF GEORGE R. CERMACK
Normal, Ill., Fire Department

A growing number of fire departments are adding specialized hazardous incident teams to existing capabilities. The Normal, Ill., Fire Department is one of them.

Normal is one of two large communities (with Bloomington) located in otherwise rural McLean County, with almost 90,000 in the two communities. Fire protection in Normal is provided by 34 career fire fighters in two fire stations. In addition to fire protection, our department also provides basic life support emergency medical service to the community.

Because of the central location of Normal in the state, several major railroads and highways intersect in the area. The potential for transportation accidents involving hazardous materials has always been a concern of our department. In 1980, we decided to take steps to expand our capabilities in this area.

Volunteers sought

A request went out to department members seeking volunteers willing to devote extra time to learning control methods of hazardous materials incidents. Six department members expressed an interest in this additional training. Initial team meetings were held during January and February.

At the same time our community was involved in annual budget hearings. We realized that our project would require fiscal support, and we took the concept of our specialized team to the town administrative staff, mayor and town council. Their response was positive and very supportive.

We realized a lot of training would be required before we could control any actual incident. We are fortunate in having a major state university in our community. Contacts were made with the chemistry department of Illinois State University, and one of their faculty members worked with our team in presenting classes in the basic chemistry of hazardous materials. Team members also devoted hundreds of hours to individual study of incident control.

Opportunities for training were taken whenever available. Team members attended several schools and seminars, and two team members attended the school instructed by Safety Systems, Inc., in Palatka, Fla. Team members also made several trips to visit people who had worked in field control of hazardous materials incidents.

We had not gone far into the planning stages when it became apparent that incident control procedures and capabilities depend on availability of necessary equipment. Since fire departments already have some of this equipment, such as protective breathing apparatus and foam capabilities, we did not have to start at point zero. We did have to add equipment such as acid suits, patching and plugging materials, nonsparking tools and other needed items.

We tried to not duplicate equipment when other area agencies had items we could get to quickly if needed. For example, we were able to locate two type B chlorine kits in our area. We purchased a type A kit and made note of where we could get the type B kits when needed. We also found that fire fighters have a lot of talent and abilities for making needed items.

Communications upgraded

Part of the equipment that had to be upgraded was our communications capabilities with various agencies that might be involved in incident control. The fire department had recently upgraded its communications system to four frequencies. In order to improve interagency communications, we had to expand the mobile radio communications network. During our first year of operations we obtained a license for a new citywide emergency radio frequency. Long-range plans call for this new frequency to be placed in every vehicle in every department in Normal to allow communications in major incidents. We also placed radio capabilities for 16 frequencies in our command vehicles and our team vehicle to expand our interagency communications.

Once we started to acquire specialized equipment, it became apparent that we needed a vehicle to carry this equipment. A decision was made to modify a 1972 reserve pumper into a specialized team vehicle. The modification of this vehicle had to be done in such a way as to retain certification of the unit as a pumper while also serving as the special team vehicle.

Town facilities and labor

The Town of Normal Public Works Department facilities were used by department personnel to complete body work on the pumper and repaint the vehicle in a high visibility chrome-yellow color. Facilities include sandblasting equipment and a commercial-size paint booth. Fire department personnel prepared the unit for painting and the public works department repainted it.

The pumper that was modified was the department's Engine 11. While looking for a name for our team, it was decided to incorporate this company into the team name, and our team became "HIT 11" (Hazardous Incident Team 11). A special uniform patch was designed and the emblem placed on the doors of the team vehicle.

The response to our efforts has been positive. We consider ourselves fortunate in that we had not had to respond to any serious incidents up to this time. We feel this is fortunate in the sense that our first year has been devoted to training and gaining basic knowledge necessary to feel comfortable working at actual incidents. We anticipate continued support from the town officials and hope to coninue developing good working relationships with other departments in the county. Future plans include the conversion of a modular ambulance into a communications and command vehicle to act as a mobile command post at major emergencies.

Our department's commitment to preparing to face hazardous materials emergencies has its associated costs, but the cost of not preparing for these emergencies is too costly to be calculated.□ □

Photo by the author

Contaminants Released at Fire Require 27 Days for Cleanup

BY JOSEPH FERGUSON

A massive explosion and fire at the four-story Berncolors-Poughkeepsie, N.Y., Dye Works last Jan. 14 killed two employees and exposed fire fighters to dangerous fumes from a possible carcinogen. Almost all fire fighting equipment and clothing was contaminated and had to be sealed in steel drums, awaiting reports from the National Institute for Occupational Safety and Health. And there was more work to do before leaving the scene.

Temperatures of 30 degrees below zero were reported during the fire attack and cleanup, with a wind chill factor of 75 degrees below zero caused by strong winds blowing off the Hudson River.

First notification to the Poughkeepsie Fire Department was a box alarm at 8:48 that Thursday morning. On arrival, Captain Clifford Kilhmire found that the front section of the 5000-square-foot building had been leveled, and the chemical-laden rubble was totally involved in fire. Second and third alarms were called immediately, bringing in city back-up companies and mutual aid from the surrounding fire districts of Arlington, Fairview, Hyde Park and Highland.

Four rescued

Poughkeepsie Lieutenant John Dakin led the rescue of three persons from the third floor of the remaining rear section of the building and one from the first floor. Two workers could not be found.

Chemical smoke poured into the air and raced southward along the Hudson, changing colors in a magnificent spectacle. Poughkeepsie Fire Chief James Davison staged fire fighting operations upwind of the blaze and ordered all men to wear full protective gear and self-contained breathing apparatus.

A 1½-inch fog line was applied to about 35 carboys of acid that were leaking and forming a cloud. An 800-gpm deluge gun was set up in front of the building to soak the rubble. Two 1¾-inch, 300-gpm hand lines were also brought in to work the debris.

At about 9:30, Davison was advised by the building owner that 400 pounds of sodium picramate, a low-level explosive used in dye manufacture, were believed to be on the fourth floor of the uninvolved rear section.

Davison immediately ordered two deluge guns directed to the fourth floor, creating a cross fire on the sodium picramate.

Evacuations begun

At this point the chief also dispatched Fire Inspector Thomas Powell to the Rip Van Winkle Tower, a high-rise apartment complex directly downwind of the fire. Experiencing burning and running eyes on the topmost floors, Powell advised immediate evacuation.

Davison ordered the evacuation be carried out under the supervision of Second Assistant Chief Thomas Armstrong, who was assisted by 50 volunteer fire fighters. The operation was coordinated with the city's police and bus departments.

Because of heavy contamination in the area, Davison could only observe the south side of the structure by venturing onto the ice of the Hudson. Assisted by Castle Point Fire Fighters Fran Diottavia and Don Stuss, he assessed the situation. From the vantage point of the ice it could be seen that the building had been split in two. Chemical drums were strewn throughout the wreckage, and the picramate drums on the fourth floor were in an exposed condition. He then ordered a third deluge gun directed at the fourth-floor containers and any that may have fallen into the rubble.

"My attitude at this point," said Davison, "was to surround and drown the fire and keep it the hell away from the fourth floor."

Two of the deluge guns were left unmanned because of their proximity to the toxic smoke and the possibility of a second explosion.

Chemicals flow into river

By 10 a.m. chemical runoff was pouring into the adjacent Fallkill Creek and the Hudson River. Davison requested the assistance of the United States Coast Guard's Marine Safety Office in Albany, which sent personnel to the scene within an hour. Port of Albany Commander P. J. Bull later called the incident "the largest chemical spill of its type in New York State history."

By 11:00 the fire had been contained to the rubble area, and fire fighters began to knock it down. A large front-end loader was brought in from the city's department of public works to get lines in underneath the debris and to aid in searching for the missing men.

The fire rekindled at about 11:30 from what Davison termed an autogeneous ignition. "It was worse than the first time," Davison said. "There were a lot of chemicals burning."

Fire fighters worked the rubble with hand lines and used the front-end loader to move debris around. "We stayed right with it, using a lot of heavy water saturation," said Davison.

Assessing contamination

By 1:30 p.m. the fire was out and fire fighters backed off to avoid further contamination. Those who showed any type of symptoms were sent immediately to Vassar Brothers Hospital, while information from the Poison Center was awaited concerning precautions that should be taken.

It was determined that nitroaniline, a potential carcinogen, was the most dangerous contaminant. Anyone who had been exposed to the smoke had to

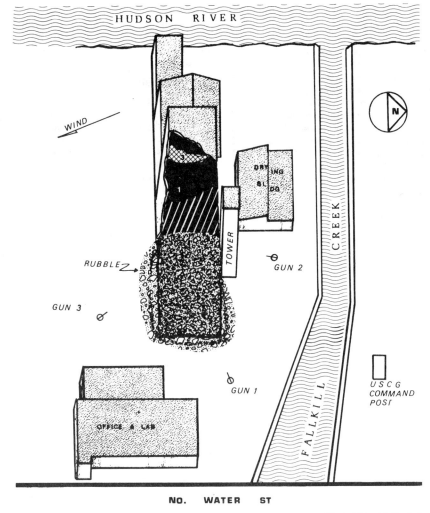

Map by Michael D. Haydock

take 15-minute showers with hot, soapy water. All clothing and gear were placed in plastic bags, and those most exposed to the fumes were sent for blood tests that evening. Tests were arranged for others at a nearby lab for the next day, with follow-up tests scheduled three to seven days later.

Fortunately, all tests would later prove negative. However, since 98 percent of the department's equipment had been contaminated and the situation was far from over, steps to obtain new turnout gear was a prime concern.

After initial inquiries, it appeared as if any new equipment would take a minimum of three months for delivery. However, Bell-Herring Suppliers of Newburgh, N.Y., was able to come up with equipment in two hours.

Four fire fighters and an engine remained on the scene overnight with police officers and Coast Guard personnel. In addition, Engine 4, located two blocks from the site, was placed on alert and staffed with four extra men and an officer.

The third deluge gun continued to saturate the rubble, while the other two were kept running into the Fallkill Creek to prevent their freezing up in the subzero weather.

Cleanup

With the Coast Guard on the scene, preparations began for the dangerous cleanup operation. Their Atlantic Strike Team was called in from Elizabeth City, N.C., and New England Pollution Control (NEPCO) of East Norwalk, Conn., experts in the field of hazardous material removal, was contacted.

NEPCO personnel, under the supervision of Vincent Brigante, the company's director of hazardous material operations, arrived at 4 p.m. the afternoon of the fire. They immediately began checking the area for airborne toxicity and explosive levels and checking the water in both the Fallkill and the Hudson. A general visual survey of the area was also conducted.

At nine o'clock that evening, Davison, together with Acting City Manager John St. Leger and other city officials, met with Brigante and Chief Warrant Officer Edward Santos of the Coast Guard to discuss strategy.

Superfund

At midnight that evening the Coast Guard set into effect the provisions of the Comprehensive Environmental Response, Compensation and Liability Act of 1980 (also known as Superfund), and assumed control of the site and supervision of all safety, fire and cleanup activities.

The Superfund (in certain circumstances) provides for reimbursement to municipalities of monies spent in the cleanup of chemical spills other than oil. This incident is one of the first times the fund may be used to such an extent, and determinations made here will affect future cases.

On the next morning, NEPCO began operations at 7:00. Overpack drums and heavy equipment were brought in for demolition and chemical removal, while staging areas were plowed of snow.

During the day the entire site was surveyed, and leaking containers of nitric, muriatic and sulfuric acid were secured. During this operation the fire department stood by, flushing the leaking carboys before the acids were pumped into the secure containers. All three deluge guns continued to operate throughout the day.

The manufacturer of the sodium picramate was contacted in order to determine the ratio of water to chemical and to find out whether crystallization due to cold could have affected its characteristics. There response was, "Handle it as you would an explosive."

"At this point we began to realize what we were up against," said Davison. It was agreed in meetings among city, Coast Guard, NEPCO and federal Environmental Protection Agency (EPA) officials, that the top priority was removal of what was determined to be eight barrels of sodium picramate from the shattered fourth floor.

Winds too strong

In spite of this agreement, the wind proved too strong that day to go ahead with removal of the picramate. This operation entailed the use of a basket-equipped crane to raise and lower two men. But they could bring down only one barrel at a time.

Because of this delay, NEPCO spent the rest of the day assembling carboys of acid that had been blown out of the building or strewn throughout the rubble. These acids, which included caustic soda, sulfuric acid and muriatic acid among others, were collected beneath what was left of the building's stair/elevator tower and washed down by fire fighters with hand lines and deluge guns. There were about 20 barrels in all.

Meanwhile, other fire fighters in class 1 protective gear, conducted a visual search of the area for traces of the two men presumed caught in the blast. NEPCO closed down operations that day at 5:30 p.m., while the fire department continued its search until 8:00. They were aided in this by lights and a generator truck from the Arlington Fire District.

All special equipment needed from outlying fire districts was obtained through the efforts of Allen Crotty, assistant coordinator for the Dutchess County Bureau of Fire, and First and Second Assistant Chiefs John Nugent and Thomas Armstrong of Poughkeepsie.

On the next morning a heavy snow was falling, but winds had died down. Preparations were made to begin removal of the sodium picramate. Fire fighters with class 1 protective gear manned two deluge guns positioned to cover the crane operator, the men on the ground and the men in the basket. A third gun was set up so that it could be swung onto the fourth floor in case of trouble.

At 9:37 a.m. 72 additional barrels of unknown materials were discovered on the floor with the picramate. These containers were also treated as explosives. In eight hours, all containers had been removed and relocated to a safe staging area.

Collapse feared

During the day the structural condition of the stair/elevator tower had deteriorated to the point where the city building inspector, Michael Haydock, stated "The collapse of the tower could occur at any moment..."

It now became critical that the 20 carboys of acids directly beneath the tower be removed immediately. Shattering of the carboys and possible mixture of the acids with other chemicals known to be on-site and in the water runoff, could result in the formation of a chlorine gas cloud and/or a hydrogen gas cloud.

Additional lights and another generator truck were brought in from the New Hackensack Fire District. The immediate area, including the Coast Guard command post, was evacuated and, because of wind direction and velocity, Davison ordered the evacuation of a 2-square-mile area to the north.

Eight NEPCO workers, in full face masks, respirators and splash suits, began removing the carboys at 7:07 p.m. Two teams of fire fighters with 1¾-inch hand lines backed up the NEPCO team, ready to knock down any vapor cloud and make rescues if necessary. One man ran the engine pump. All fire department personnel involved, including Davison and Haydock, who supervised operations, wore full protective gear.

Poughkeepsie Mayor Thomas Aposporos presented medals of honor on April 16 to Lieutenant John Dakin, Fire Fighters William Burgin and Arthur Ghee, Fire Inspector Thomas Powell, and two police officers for entering the third floor of the building after the explosion to rescue three Berncolors employees "without hesitation or regard for their own safety or well-being, and with the immediate danger of a secondary explosion and further building collapse," according to Fire Chief James Davison.

All others who participated in the fire fighting and spill control received exceptional service awards and certificates of appreciation. Davison awarded Bull and Brigante the rank of honorary fire chief.

All 20 containers had been successfully removed within 15 minutes. A meeting was held among all officials involved, and it was agreed that conditions were now safe enough to allow the evacuees from Rip Van Winkle, as well as those moved out of the north sector, to return home. Only the residents of three houses directly across the street remained away until Tuesday.

At this point, overnight fire personnel was cut back to two. Extra manpower at Engine 4 was maintained.

The next day, Sunday, the tower was demolished and the search for the bodies continued.

Next crisis: ice jam

At about 3:00 that afternoon, Davison was informed that an ice jam was forming and beginning to dam the Fallkill Creek about a quarter-mile above the site.

Fearing a flash flood in the event of a sudden breakthrough of backed-up water, the chief requested the assistance of Leroy Fine of Dutchess County Flood Control. Fine supervised operations as fire fighters worked to clear the jam using 1¾-inch lines to cut slots through the soft, still-forming ice. Later the jam was cleared completely by the Dutchess County Highway Department.

Operations at the site shut down early that day due to the 20 to 25-knot winds with heavier gusts.

While the picramate was being removed on Saturday, a decision was made by the building inspector concerning the adjacent laboratory and office structure which proved its value two days later.

The building, though not directly involved in the fire and explosion, had had all its windows blown out by the force of the blast. Upon inspection by Haydock and Powell, it was found that the entire inside had been disrupted from both the concussion and subsequent exposure to the elements. On the basis of this inspection, Haydock ordered that no one but NEPCO personnel be allowed into the building.

More explosives, cyanide

On Monday morning, NEPCO workers discovered a 6-ounce container of crystallized picric acid, a high-level explosive, next to containers of sodium cyanide and potassium cyanide, both highly poisonous salts. Roy Gould of NEPCO volunteered to remove the picric acid to a snow bank in a safe area. The cyanides were taken to Albany by a state environmental official.

After consultations with chemists from Texaco and Vassar College, who described the substance as "a potential time bomb," it was decided to detonate the picric acid on-site.

The U.S. Navy Explosive Ordnance Disposal Detachment at Colts Neck, N.J., was contacted, and a five-man team commanded by Lieutenant Joseph Tenaglia arrived at the site at about 6 p.m. A plastic explosive charge was secured to the acid container and it was safely detonated at 7:30.

The painstaking cleanup and search for the bodies continued over the next several days. Rubble had to be carefully sorted for signs of the two presumed victims and to make determinations on the contents of several hundred drums strewn throughout the wreckage. In addition, all material removed had to be sorted into three piles: contaminated, mildly contaminated and not contaminated. These determinations were made by EPA and Coast Guard officials.

Bodies found

The bodies of the two missing company employees were found on Jan. 25, 11 days after the explosion. Fire department apparatus and personnel remained until Feb. 8. Control of the property was returned to the owner on Feb. 10.

The entire operation had lasted just a few days short of a month. During that time 22 million gallons of water were pumped by the fire department; turnout gear, hoses and various other equipment were contaminated or destroyed; and the total city cost of the operation was $155,000. How much of that the City of Poughkeepsie and the fire department will be reimbursed under the Superfund remains to be seen.

The incident at Berncolors also pointed out the necessity of a new commercial inspection program that the building and fire inspectors had been working on prior to the explosion. Implementation of this system was sped up and has since been placed into effect. ☐ ☐

AMMONIA LEAK

With industry assistance, Morton, Ill., fire fighters trained specifically for an ammonia leak. Then a real emergency occurred.

Fire fighters use hose stream to control gas cloud while moving in on broken valve.

BY GARY BLACKBURN

In enclosed protective suits, Cloyd and Arnold work on the valve—photos by John Cary.

The local Amoco fertilizer plant in Morton, a town of 15,000, provided the ammonia for the training sessions, discharging enough to create a small cloud. Fire fighters donned protective chemical suits and went in. They gained the experience and confidence to handle any spill, according to Fire Chief Dick Campbell.

Then last May 5, a truck backed into a feeder valve on the Amoco plant's main 12,000-gallon ammonia tank. The valve broke, permitting a white cloud of super-cold and noxious gas to escape into the air.

Amoco employees called for assistance at 6:06 p.m. on that Wednesday. It may have been about the ideal time for such an emergency. Fourteen members of the fire department were already at the fire station, located about seven blocks from the leak site. It was election night for a new chief and officers. And the meeting was just about to start.

Weather was about as cooperative as it could be, too. Wind direction was from the southeast, blowing the growing ammonia cloud away from populated areas and a major interstate highway, I-74. The site had been chosen by Amoco because it was on the north side of town, where prevailing winds would usually blow any leaks away from populated areas. Temperature was a moderate 72 degrees and it was cloudy. Not bad conditions for fire fighters to work in air packs and closed chemical suits.

Two pumpers and a rescue truck arrived at 6:10 p.m. Evacuation was considered because a new residential subdivision was within one block of the leak site. Calm winds blowing away from the site, however, made evacuation unnecessary.

The next consideration was traffic and crowd control. Ammonia fumes can freeze the unprotected skin up close and cause lung and eye damage even at a distance. At least two persons driving past the site on nearby Main St. were overcome by the fumes. They ran off the road, but no one required hospitalization or serious treatment.

Units of the Morton Police and Morton Emergency Services and Disaster Agency were called as soon as fire officers saw the nature of the emergency. They set up a safe zone and blocked traffic while fire fighters worked on the leak.

Fire officers quickly determined no one was injured or close to the leaking fumes. Two fire fighters, Bob Cloyd and Jim Arnold, put on SCBA and enclosed protective suits.

The leak should have been easily stopped by following the line back to the main tank where another shutoff valve was located. That valve malfunctioned, however, and did not shut down the flow.

Meanwhile, four other fire fighters using SCBA set up a cross stream with each team of two using a fog nozzle on the end of a 1½-inch line. They were supplied by a new 1000-gpm pumper with a 2500-gallon tank. Another support truck was moved into position to provide more water.

The cross-stream fog screen did a lot to dilute the escaping ammonia and prevent the cloud from spreading. Meanwhile, the two fire fighters in chemical suits moved in to assess the damaged valve.

It was broken and had to be replaced or plugged. Luckily the fertilizer plant manager had another standard-size valve on hand in the office. It was located and given to the men in the chemical suits. They removed the damaged valve, installed the replacement and shut down the leak.

Since that leak, the fire department has acquired some tapered wooden plugs that would have quickly jammed into the fire opening to plug the low-pressure leak, had the replacement valve not been available.

The fire department estimated that between 200 and 300 gallons of ammonia was lost. The entire incident was complete and trucks were back in the station by 7:10 p.m., just an hour after arrival.

During training sessions with the ammonia, the men always work in pairs. "You learn to trust the person you're with. It really helps the men be closer, feel good about their job and capable," Campbell said.

The department has four chemical suits and drills with them regularly. Each time they are checked for leaks or tears that might mean injury to those inside.

Campbell also warns of operating equipment around ammonia. The ammonia can break down the oil inside truck engines and pumps and cause them to run hot or freeze up very quickly. So don't drive through the ammonia cloud or position a truck where the escaping fumes might engulf it.

Training materials for the Morton department, now celebrating its 100-year anniversary, are obtained from the International Fire Service Training Association. And drills are held every Tuesday night on some aspect of training.

The department recently obtained A, B and C kits to handle chlorine leaks. A is for 150-pound vessels, B is for 1-ton and C is for rail car or transport repair.

A task force made up of men within the department is being established to handle any leak at a swimming pool or water or wastewater plant. ❑ ❑

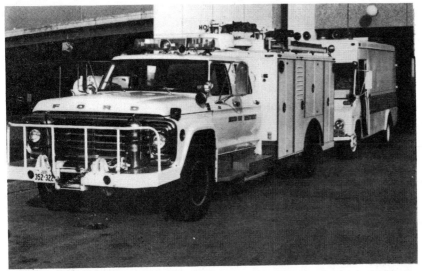

Apparatus used by Houston Fire Department hazardous materials team consists of a 1979 heavy rescue unit, R-1, with a crew cab and a 1967 van, HM-1.

Storage walls inside HM-1 provide space for tools and emergency devices.

Houston F.D. Haz-Mat Team Finds Activity Rises as Reputation Spreads

BY R. L. NAILEN
Staff Correspondent

"Energy Capital of the World"— that's Houston, the nation's leading center of petroleum refining and chemical manufacture, as well as the third busiest seaport.

The 40-mile Ship Channel connecting the city with the Gulf Coast, used by 5000 vessels yearly, is lined with petrochemical plants. So when Houston Fire Chief V. E. Rogers heard an International Association of Fire Chiefs presentation several years ago on hazardous material training in the Jacksonville, Fla., Fire Department, it's not surprising that he decided Houston should equip itself to deal with chemical hazards.

In 1978, he assigned 10th District Chief Max McRae to develop a response team for chemical emergencies. That team completed its first year of service last Oct. 5 with an already outstanding record of saving lives and property while sustaining no on-duty injuries.

Six men on each shift

A crew of six on each of three shifts operates the team's two vehicles: Rescue 1 (R-1) with five men, plus Hazardous Material 1 (HM-1) with a driver. Whenever fire dispatchers have reason to believe a chemical hazard exists, both units are dispatched together. On routine rescue calls, R-1 goes alone.

Conventional emergencies for HM-1, such as gasoline tanker spills, are quite rare. In fact, many of the team's runs involve toxic or corrosive—but nonflammable—materials.

"If it's made anywhere at all, it's probably made here," said acting Captain Bill Hand of R-1, speaking of the variety of chemicals produced and transported in the Houston area.

"We especially encounter a lot of dangerous intermediates," he continued. "These are chemicals not intended to leave a plant for sale, but produced as some intermediate step in a process generating a different product, such as a 'hot' pesticide.

"Houston is a major shipping point on the Southern Pacific Railroad, also. They tell us just about every major hazardous chemical shipment on their system either originates or terminates here."

Although transportation accidents are significant, McRae pointed out that "transportation isn't necessarily going to be the problem. We have more fixed system accidents, as in clandestine drug labs, than we do on the road."

Responses increase

The team's responses were at first infrequent until industry and other agencies became fully aware of HM-1's expertise and availability. There were only 28 calls for HM-1 in 1979, but the rate is now 30 a month, with a total of 300 for the first year. About half the calls actually require the team's services.

Sometimes the crew has hardly had a chance to learn a technique before being called on to use it in an emergency.

"We had a workshop on chlorine kits," explained McRae. "A chemical company brought us in a 150-pound cylinder for practice. Right while that was going on, the on-duty team members got a run and had to put a side patch on a leaking chlorine cylinder."

They've done their share of things that "couldn't be done."

Chemical incident

Hand recalled, "One of the two chemicals we were told by Dow to leave strictly alone is anhydrous hydrogen chloride. But, sure enough, we got a call to the docks where a truck-mounted 20-foot cylinder, being shipped to Australia, was leaking that stuff. A forklift driver had mistakenly stuck a fork through the cap while trying to pick the tank up. The acid vapor cloud inside the warehouse was so thick we couldn't tell where the leak was. There were 300 drums of export-only pesticide around with the possibility of a major spill into the Ship Channel. So we had to do something.

"We found a similar tank elsewhere in the port area, then studied it carefully to figure out where the leak probably was and how to stop it. Then we were able to do the job."

This past May, HM-1 was called to a process plant where sulfur dioxide was escaping from a 200,000-gallon tank of spent sulfuric acid. This was acid which had been used in various reactions, so it contained a variety of compounds of uncertain properties. In this case, the plant chemist had no idea what reactions might be taking place within the tank.

Thermal layers evident

"There were obvious thermal layers, though," said Hand, "because at various levels around the outside of the tank you could see the paint blistering from the heat inside. We got lines on it to cool things off.

"What had happened was that an internal explosion had pushed the tank roof up and torn it loose at the edge. We had a rip 2 feet wide by 35 feet long near the top. We had to go up there in a Snorkel basket and cover the tear with salvage covers to stop the leak."

The 15 members of Houston's hazardous materials team were trained during a three-week period in September and October 1979, using a 104-hour program (see chart) worked out by McRae in consultation with other officers over a year's time. About half the training was hands-on, the rest classroom work.

"Most of the initial instruction was offered by people from industry emergency response teams," McRae explained. "The 16 hours of basic chemistry was given by our own fire protection engineer, who is a community college instructor in the subject."

Six basic functions

Department members installed storage shelves in the HM-1 vehicle, formerly a rescue van. Equipment and tools originally put on board were intended to carry out these six basic functions (the principal tool involved appears in parentheses):

1. Measure explosion hazard (explosive vapor detector),
2. Neutralize acids (soda ash),
3. Emulsify hydrocarbons (Misco "No-Flash" compound),
4. Stop leaks (clamps and plugs),
5. Contain spills (dike materials), and
6. Handle class B fires (foam).

As team members gained experience, additional items—some of them unique—have been added.

For example, following trouble sealing off a leaking manhole in a 175,000-barrel tank after a fire, a set of large wrenches was requested. Both pipe and socket wrenches in a wide range of sizes are now carried by HM-1.

Variety of stoppers

There's a complete kit of rubber and tapered wood stoppers. Their size indicates the fact that, however dangerous leaks may be, most of those encountered are fairly small.

Dealing with a cracked or torn chemical drum is a special problem. For that, team members developed their own external "strap clamps" which can be locked in place outside a drum or tank wall without access to the interior.

The T-bolts work like the toggle bolts a householder might use to attach furnishings to wallboard. After slipping one end through the tear in the drum, the team member turns the bolt a quarter turn to pull it tight, so that the

---

1st Training Program For Haz-Mat Team

		Days
1.	NFPA program, "Handling Hazardous Materials Transportation Emergencies"	2
2.	Field work with alkyls, chlorine, corrosives, LPG, refinery operations, tank cards, patching & capping containers (at various chemical plants	5½
3.	Basic chemistry	2
4.	Self-contained breathing apparatus	½
5.	Response team operations	1
6.	Foam	½
7.	Cryogens	½
8.	Tank truck and loading rack regulations	½
9.	Pesticide spills and fires	½
10	Care & use of explosive vapor detector, CO meter, and acid suits	½

---

gasketed clamp strap is compressed over the opening. Because it may then be necessary to relieve internal pressure, or even to transfer the contents to another container, the clamp is fitted with a pipe connection which can either be plugged or used as a drain.

Some team equipment is kept on R-1.

Between them, the two vehicles also carry:

1. A complete chemical library. Reference books include: NFPA "Hazardous Materials Guide," "Firefighters Handbook of Hazardous Materials," "Farm Chemical Handbook," "GATX Tank Car Manual," "Recognition and Management of Pesticide Poisonings," and a number of others.

2. Five acid suits. Here, too, local conditions are bringing some variations. The team is now experimenting with an internal air-cooling system for the suits, supplied by 300 feet of hose from a cascade system.

"In this climate," Hand explained, "we try to carry 5 gallons of ice water to cool off with, but we still can't usually work more than 15 minutes in a suit."

Also, the stretchable O-rings used to anchor gloves to the suit wristlets had a habit of slipping off during use. The HM-1 crew has replaced them with plastic cable tie-straps locked to a fixed size. These will not stretch.

3. Three chlorine kits.
4. Eductors, nozzles, and generators for protein, high expansion and AFFF foams.
5. Tank car adapters and wrenches.
6. Gasket sealants and adhesives plus leather and neoprene gasket material.
7. Dry sand and soda ash (300 pounds each) plus 80 gallons of various foam concentrates.
8. Epoxy patching kits plus glass tape and cloth.

Clamps and plugs are among the emergency devices carried by HM-1. In the foreground are several types of clamp straps along with several shutoff clamps for small pipes.

9. Emergency personnel washdown kit.

In addition, R-1 carries a full complement of conventional heavy rescue gear: lifelines, floodlights, chain slings, jacks, hoists, Hurst Tool, radiation monitoring kits, emergency medical supplies etc.—more than 350 items.

Reasons for location

Asked about the team's downtown Fire Headquarters location, often a long distance from the serious petrochemical hazards, Hand replied, "We're needed everywhere. So this central location is the best. It does take us 15 minutes to respond to some areas. But it can take three or four hours for an industrial team to get into action by the time all the manpower and equipment is assembled.

"Besides, the property risk may be greater along the Channel. But in the southwest and northwest parts of the city, which are the fastest growing with lots of big apartment developments, the life hazard is much greater. So we have to be able to reach those areas quickly too. And we've had some runs right at our back door."

It's possible that as the city continues to grow (an estimated 5000 more residents arrive each month), a second team will have to be established elsewhere.

The action plan on which HM-1's operations are based sets these priorities:

1. Safety of citizens.
2. Safety of fire fighters.
3. Area evacuation when necessary.
4. Control of the situation.
5. Stabilization of the hazardous materials, and/or . . .
6. Their disposal or removal.

Cleanup capability needed

Commented McRae, "The original intent was for us just to stabilize the situation, then let industry handle the cleanup or off-loading operation. But we found out pretty quickly that it just doesn't work that way. Often you can't get hold of the people, especially at night. If you hang around waiting for someone, your team is tied up. If you leave, you risk a blowup for which you will be blamed. So we must have the capability to clean up or off-load ourselves.

"And the commercial cleanup firms don't always have the expertise needed. They may not bring grounding cables, for instance, so we are getting our own. We don't want to blow up along with them."

One of the disposal techniques learned by HM-1 personnel is the use of "recovery drums." These are oversize containers into which a leaking chemical drum can be placed, the lid closed, and the whole works safely transported whenever desired.

Because of their close ties with the chemical industry, HM-1 has responded outside Houston, in one instance as far away as Texas City (30 miles).

"We work with CIMA, the Channel Industries Mutual Aid group," McRae explained.

That association now has 80 member firms in the paper, drug, petroleum, chemical, and transportation industries.

First District Chief W. M. Whately said it was at first feared the team would not get enough runs to justify its existence, but responses have since occurred far more often than expected.

"Not only did we not realize how active we'd be," added McRae, "we also found we're on the scene longer than the typical engine company, for instance. The team is often tied up for several hours, even half a shift, instead of just spending a half hour to clean things up and return to quarters."

The 43 different materials encountered by HM-1 in its first six months alone read like a chemical dictionary. They are explosive, toxic, corrosive—even radioactive. To give only a partial list:

Sodium cyanide, methyl methacrylate, mercaptan, cyclohexane, nickel carbonyl, hydrogen peroxide, monoethylaniline, iridium 192, aromatic naphtha, sulfur trioxide.

Ownership is problem

They come in tanks, drums, cylinders—every kind of container. Identification is not a common problem. "CHEMTREC has been a tremendous help to us there," McRae said. "The trouble is in finding out who owns the stuff, especially around the port. You can't always identify the responsible carrier. It's much simpler on the highway or the railroad."

There have been some public relations benefit for the fire department from the activities of HM-1, although the general public understandably has little contact with the team.

Explained Hand, "We did get TV coverage on Channel 13 after only a week in service. It was an intermodal container in a trucking yard, with a supposedly empty hydrofluoric acid tank. Somebody had removed the safety valve and left the hatch cover open, which was messed up so it wouldn't close properly anyway. But there were about 30 gallons inside. We went to a truck stop for some old inner tubes, then got some scrap plywood to make a new hatch cover—chained it down over the inner tubes with a hydraulic jack, then put rubber plugs in the relief valve hole. We didn't have much of our own equipment then and had to innovate.

Manager grateful

"And we got a lot of gratitude from an apartment manager. A building owner had used some two-part insulating foam on the roof, then left 14 drums of the leftover isocyanate sitting outside the laundry room. After heating and cooling for three years, the seven drums still containing chemicals began to leak into the parking lot. You could still read the labeling, so we reached the manufacturer through CHEMTREC and were able to take care of disposal. It would have cost the manager $100 a drum to get that done privately."

The team's education has of course not ended with the initial program. From time to time, developments are explained. Now on trial is an acid sump cover for bottom outlet tank cars. A new sealant is available, both for patching and for forming a layer on sand dikes to seal their surfaces. A local industry recently donated a tank car to the fire department for mounting at the training academy. A variety of domes is now being sought for use with it.

A few team members have had other related training. Two have some college chemistry background. Three spent a week at Texas A & M fighting hazardous material fires and working on diking.

Seminar held

Most recently, McRae led a three-day September seminar to introduce new material. Among the items covered was the Southern Pacific's "dome mobile" (Fire Engineering, January 1980) program. The Coast Guard, which has some jurisdiction over port activities along the Ship Channel, came in to explain its port safety work. There was a session on pesticides and another on high pressure gas cylinders.

Added McRae, "We also have occasional critiques when the men of all three shifts can get together and discuss common problems. It's the only chance they have to compare notes."

For the future, refresher training will continue. And, whether or not a second HM unit goes into service, a backup supply of personnel will be needed.

Said McRae, "Nobody is 'burned out' yet, but we do need more people. Vacations and holidays tend to make things tight."

Meanwhile, Captain J. E. Knoll has spent several months (for all three shifts) presenting a three-hour program to the entire fire department on the hazardous material problem and the operations of HM-1, which are certain to increase as industrial chemistry grows more exotic. ☐ ☐

Entire Buildings Are Labeled With NFPA 704M Marking System

BY BRADLEY ANDERSON
*Fire Inspector
Charlotte, N.C., Fire Department*

A hazardous materials marking system for buildings is providing Charlotte, N.C., fire fighters with extra important information about dangerous materials that may be present in a structural fire.

Some fire departments already maintain a central building record system, but that system puts extra work on fire alarm operators when they are most heavily involved in dispatching multiple companies or receiving calls. And a central file cannot easily be referenced if smoke is reported on a certain street but the exact address is not known.

Need tragically demonstrated

The need for immediate identification of the presence of hazardous materials was tragically demonstrated in Charlotte in 1959. Fire fighters responding to a fire in an abandoned chemical plant being demolished found a vat of burning kerosine. They were unaware of a submerged 100-pound block of metallic sodium that had reacted with rainwater and ignited the kerosine after the building's roof was removed. When fire fighters directed a fog pattern into the vat, an explosion occurred that nearly killed one fire fighter and injured several others. Pieces of sodium burned the paint off cars two blocks away. Had the building's roof been in place, 10 or more fire fighters in the building might have been killed.

In 1960, the fire department began supplementing a central record of each building's contents with warning signs posted on the exterior wall of buildings and tanks containing hazardous materials. These signs provided immediate warning of the presence of hazardous materials so that special caution would be exercised by fire fighters operating on the premises.

The signs used then were simple: They were based on the ICC shipping labels in effect at the time and consisted of a 24-inch-square diamond painted red, green, yellow or white (depending upon the nature of the hazard) with a verbal warning such as "flammable,"

Arriving fire fighters can quickly tell the nature of any hazardous materials on the inside by looking at the color and number-coded placards on the outside.

"combustible" or "acid." Fire inspectors requested businesses storing large quantities of hazardous materials to post the signs. Businesses complied on a voluntary basis; there was no such requirement in the fire code.

New system description

Charlotte adopted NFPA's 704M Hazardous Materials Marking System in place of the older marking system in 1974. Blue, red and yellow diamond signs each contain a number from zero to four indicating the relative severity of the health, flammability and detonation hazard, while a white diamond indicates special considerations such as the presence of water-reactive substances, oxidizers or radioactive materials. The 704M system provides a simple means of conveying comprehensive information at a glance.

Businesses voluntarily converted to the new marking system, realizing the value of the information that it gives fire fighters.

In 1977, the city council adopted NFPA 1, Fire Prevention Code, as the official city fire code. This code replaced the previously used AIA code. Under paragraph 3-9.4 of the new code, posting of warning signs such as 704M is now totally enforceable, although legal action to gain compliance has never been necessary.

All businesses inspected

All businesses using significant quantities of hazardous materials are inspected every six months by the fire prevention bureau. Fire fighting companies are responsible for in-service inspections in less complicated structures while the fire prevention bureau handles the more complex inspections. Company officers refer inspections of hazardous materials that they discover to the bureau. A fire inspector then visits the premises, ensures proper storage and handling of the materials, prescribes the appropriate type and location of 704M placards and, upon compliance, issues a permit. Information on the type, quantity and location of hazardous materials is recorded on a building record which is sent to the fire alarm office. The building record also shows the 704M placarding of that occupancy.

During subsequent six-months visits, the inspector ensures that newly added hazardous materials are stored and handled in accordance with the code, updates information on the building record, prescribes any related changes

The placard for a building shows cumulative ratings. For example, sulfuric acid (left) and paints (center) stored in a building call for a combined placard (right) featuring the highest number in each category.

in the 704M signs and updates the permit.

Showing degree of hazard

Most of the materials found during the inspection can be referenced in the NFPA's "Hazardous Materials Guide" to determine the appropriate placards. Often there are several different materials stored in the building, each calling for a specific placard combination. In these cases the most severe degree of hazard presented by each material is posed for each category on the sign. For example, a garage that provides a variety of services may stock battery acid and also have a supply of paints on hand. The sulfuric acid would be placarded with a blue health hazard marking of 3, no red flammability marking, a yellow reactivity warning of 2 and a special warning advising careful use of water. The paints call for a blue 2 and red 3.

The inspector combines the categories into the highest degrees of hazard of each and in this case instructs the business manager to post a blue 3, red 3, yellow 2 and white W.

Individual category signs are posted for each hazard in a visible area on the exterior of the building so that fire fighters responding to an incident there will spot the signs as they arrive. In some buildings we require 704M signs in storage rooms and laboratories as well as on an exterior wall. Storage tanks containing hazardous materials also have appropriate 704M diamonds.

Fireground Application

For any alarm of a building fire fire alarm dispatchers still check the address given against the building record file. In addition to relaying hydrant locations obtained from the building record, hazardous material information is also broadcast. But if the exact address is not known, the first-arriving company checks for 704M signs as a part of the initial size-up. When possible, the fire company relays the exact address to the alarm office and gets more precise information, such as the exact type of materials and their location within the building, before an overly aggressive attack is mounted.

If the situation requires immediate action and time cannot be taken to reference an address, the incident commander still has some basic information about the nature of the hazard from the building's mounted diamond placards. The diamonds tell at a glance whether special protective equipment is necessary (blue category), whether to anticipate a sudden increase in the magnitude of the fire as well as what may be burning (red category), whether an explosion is likely to occur (yellow) and if the use of water should be restricted or other special considerations be given to the contents of the building (white).

In the 22 years that the placarding of signs has been requested and—more recently—required, no problems have been encountered gaining the compliance of building owners or managers. During initial inspection of hazardous materials, the fire inspector explains the purpose of the signs to the occupant and draws a sketch of the prescribed placard. Occupants may make their own placards or they may order the placards from any sign painter in the city.

Resolving problems

One occasional problem is that hazard categories are posted in the wrong position relative to each other. When this occurs, the inspector makes sure that the colors relate correctly to the hazard, which accomplishes the objective of the sign. No corrective action is generally required.

To make it easier for the occupant to comply with the 704M requirements in businesses where materials do not present all four categories of hazard, only those hazardous categories present are required to be posted. If in a dry cleaning establishment the only hazardous material present in large quantities is perchloroethylene—which presents a health hazard but no fire or reactive potential and needs no other special consideration—only a blue diamond exhibiting the number 3 is required to be posted instead of all the others with a zero. One criticism of this policy is that if a part of the sign falls off the building, fire fighters will misread the hazard. This concern is valid, but such a condition never exists for an extended period of time because 704M signs are checked each six months to ensure they accurately identify the hazards of existing and new materials.

We also try to be reasonable with respect to the size of signs. Category diamonds smaller than 24 inches square are permitted where they are posted on a door or in a space that will not easily accommodate a full-scale sign. Aesthetics are also occasionally allowed to influence the size of the diamonds where this will pose no difficulty in spotting or reading the 704M warning.

Through a comprehensive building record file and placarding system, our department has reduced the potential for fire fighter casualties resulting from hazardous materials. The system has been quite effective; no Charlotte fire fighter has been injured by stored hazardous materials since the program was initiated after the 1959 tragedy. This identification system provides a means of distinguishing a routine building fire from a hazardous materials incident before it is too late. With increasing numbers of hazardous materials being introduced daily, such a system is worth considering as an integral part of any fire inspection program. □ □

Photos courtesy of Southern Pacific Company

Dome mobile, which contains four types of tank car domes, is used by Bob Andre, second from right, to explain to Oakland, Calif., fire fighters how to handle leaks. Carbon dioxide cylinder to pressurize simulated leaks, is in box at rear of trailer.

Railroad Goes to Fire Departments With Hazardous Materials Program

BY R. L. NAILEN
Staff Correspondent

A century ago, railroad fire hazards were simple. An occasional locomotive spark might ignite a town's wooden sidewalks or burn down the livery stable.

Today's railroad fire and rescue problems are far more complex, and the tank car loaded with exotic chemicals is only one of many hazards. To deal with these problems along the right-of-way, a number of major railroads, particularly in the South and West, have formed their own high-level corporate divisions (examples are the Burlington Northern, Union Pacific, and St. Louis-San Francisco systems). Such a group typically has three objectives:

1. Educating fire fighters on the nature of railroad fire problems and how to attack them.

2. Providing specialized emergency equipment which may not otherwise be available to many fire departments.

3. Making trained manpower available to assist fire departments at an emergency in every possible way.

Railroad president involved

A leader in this field since 1975 has been the giant Southern Pacific Transportation Company, descendant of the West's first transcontinental line, operating over 18,000 track miles in a dozen states from Missouri to Oregon. Southern Pacific's President D.K. McNear is a member of the 12-man Task Force on Rail Transportation of Hazardous Materials that was put together a couple of years ago by the Manufacturing Chemists Association and the Association of American Railroads (AAR).

Each year, an estimated million carloads of hazardous materials move over United States railroads. The SP handles 120,000 of them, or more than 10 percent. From San Francisco, Bob Andre, the SP's hazardous materials control superintendent, heads a group of six men located in Houston, Los Angeles, and Roseville, Calif. Altogether, this team has nearly a century of experience in its field. Its services are available 24 hours a day.

One of the group's first tasks was to develop training programs on rail emergencies that could be used to teach both employees and the fire services along the SP's routes. Mention the problem of hazardous rail cargoes to the average fire fighter and he at once thinks of the spectacular train wreck, the explosion which wipes out a whole community.

Leaking tank cars

Despite their destructive power, however, such accidents are infrequent. Far more widely encountered is the leaking tank car. Often unsuspected, requiring specialized knowledge to repair, such leaks occur as often as 400 times a year along the Southern Pacific system alone. So this problem was the first one dealt with by the hazardous materials control group.

"This is one of the most common headaches we face," explained Andre.

"Very few people realize that railroads own none of the compressed gas tank cars, the kind that frequently find their way into newspaper headlines. Nor do railroad personnel load or unload these cars. Our leakers are inherited as a result of improper loading or unloading procedures by shippers. So our employees were unfamiliar with leaking tank cars and yet were being faced with them frequently."

Therefore, the SP produced a unique audiovisual training program titled "Keeping Hazardous Materials Contained in Tank Cars." It includes 79 slides accompanied by narration. But looking and listening are not enough.

"We felt that nothing short of actual tank car equipment could provide the all-important first time, hands-on experience of actually repairing a leak," Andre pointed out.

Dome mobiles built

To give fire fighters a chance to actually handle typical tank car valve equipment, the company has constructed four complete replicas of tank car domes. For each one, three different cars were stripped of their domes, valves, and appurtenances, and this equipment was mounted on a 15-foot trailer. These trailers, called "dome mobiles," have been used to train 20,000 fire fighters in 500 fire departments along SP lines since 1975 (besides 3000 railroad workers and many shipper representatives).

The tank car dome, a circular steel housing at the car's top center, houses a complex array of fittings that vary considerably among the three main car types (compressed gas, flammable liquid, and acid). A typical compressed

Fire fighters demonstrate indirect method of attacking boxcar fire by operating a fog nozzle through hole opened in roof. Car is laddered well away from closed doors.

gas car dome, for example, contains two liquid eduction valves, a vapor eduction valve, a sampling valve, a spring-loaded safety relief valve, a level gaging device, and a thermometer well. These cars are loaded by volume, which will vary with temperature, so a check of car loading must include measurement of contents temperature.

To add realism, the trailer's compressed gas fittings are arranged to simulate leaks that can be seen and smelled. A mixture of odorized carbon dioxide and Freon is fed to the fittings under pressure.

Explained Andre, "When this stuff leaks, there is absolutely no discernible difference between the simulated experience and reality. After thorough instruction in repair procedures, leaks are produced by the instructor for correction by the trainees."

Subjects covered

Participants become familiar with the special tools needed, such as T-handle and crowfoot wrenches. They learn about rupture discs that relieve internal pressure in acid cars. They learn the difference between top-operating and bottom-unloading valves. Some of the other techniques covered are:

1. Stopping packing leaks on eduction valves.
2. Replacing 0-ring seals on safety valves.
3. Handling outlet cap leaks on bottom-unloading flammable liquid cars.
4. Replacing cover plate gaskets on acid cars.
5. Workings of the gaging device, which determines the level of the liquid-vapor interface inside a compressed gas car. It is here where most leaks occur.

The four dome mobiles are based in San Francisco, Los Angeles, Houston, and Pine Bluff, Ark. A complete training session with the unit takes 1½ hours.

Locomotive fires

Dangerous though they may be, tank car leaks are just one of the problems for fire departments along a railroad line. During the 1970s, the nation's railroads have annually suffered $15 to $20 million in fire loss. Fires in locomotives and freight cars present several unusual hazards calling for special fire fighter training.

For that, the SP produced a second audiovisual program titled, "Emergency Fire Fighting Procedures on Railroad Equipment." This one-hour presentation includes 56 slides covering two subjects:

1. Potential fire problems of diesel-electric locomotive.
2. Fires in closed boxcars, or in the two types of containerized freight: truck-trailers on flatcars, or the intercontinental shipping container.

In the first category, the training highlights Class B and C locomotive fire dangers. A typical power unit—and it may take five or six of them to pull a long train on inter-city main lines—carries a 2500-kw main generator, a

Oregon fire fighters watch instructor demonstrate top-operating mechanism found on some tank cars for controlling bottom-unloading valve.

27

74-volt starting battery, and a 4000-gallon tank of diesel fuel. The SP's program shows the location and use of emergency shutdown switches (both inside and outside the locomotive) to stop the generator or to disconnect the battery, how to gauge the amount of fuel on board, and where fire extinguishers are located.

Training put to use

This education was timely for the Elko, Nev., Fire Department. On the evening of Jan. 5, 1978, less than a month after Elko fire fighters took the training, a freight train rolled into town with a locomotive on fire.

Chief Bill Fogle later told SP officials, "The captain on shift knew just what to do to shut down the unit and bring the fire under control. Afterwards, I was talking with the crew on the train and they were relating to me about the firemen knowing what to do and that they were glad that they did. The excitement caused the train crew members themselves to forget just what to do. So thanks for a top quality training package."

Incidentally, although SP tracks pass through Elko, this incident involved equipment of another railroad. Diesel locomotive operating features are much the same throughout the industry—everywhere in the country.

Only a few weeks later, Elko began working with the dome mobile program—and had a tank car incident in the city at the same time.

Refrigerator cars

Similar to the handling of locomotive fires, though on a far smaller scale, are problems with the mechanical refrigeration equipment in modern refrigerator cars. Unlike the old type, kept cool by blocks of ice stacked in car bunkers, these cars contain a complete electrical cooling system powered by a 220-volt generator. This is driven by a diesel engine in a large side compartment near one end of the car. Beneath the car is a 500-gallon fuel tank. These cars also have emergency shutdown controls, the location and use of which are shown in the SP training package. Some piggyback truck trailers shipped on rail flatcars use similar engine-driven refrigeration systems.

It is the boxcar fire, however, that presents the most common rail fire problem. A blaze within a closed container shipment is of the same type. Accustomed to attack fires within closed spaces by first ventilating, then extinguishing, fire fighters will find such tactics unsuited to boxcar fires.

Instead, Southern Pacific teaches the indirect method. Since its development by the company in 1964 to fight baled cotton fires, this procedure, according to the SP, "has since proven to be one of the safest and most efficient approaches to fighting all types of freight container fires."

Indirect attack used

Key to the indirect method is leaving car doors tightly closed to keep oxygen from the fire within. Another reason for the closed-door policy is the likelihood that the car will be loaded to the roof. That means fire stream access through opened doors may be quite limited. Moreover, burning debris falling through the doorway may endanger personnel.

Rather, fire fighters should find the general location of the fire by feel or by observing where externally-applied water dries quickly against the car wall. Next, they should cut open the car roof over that area, then introduce fog, swinging the nozzle from side to side until no visible smoke exists from the roof openings. Then the car doors should stay closed for at least another hour to ensure final extinguishment.

Fire fighters used to an "open it up, then knock it down" technique may not be aware of the risk of explosion inside boxcars containing no hazardous or explosive commodity. This can occur with auto tire shipments, for example. Opening the car doors, or making vent openings in the car bottom as well as the top, can let in enough air to violently ignite gases produced by long smoldering combustion. Furthermore, applying water to the contents can build up a high steam pressure within the car.

Any internal explosion is likely to blow off the doors, which form the weakest areas of the car structure. Therefore, everybody should keep well away from them.

Trainees are also cautioned that burning non-hazardous materials often create toxic gases, so the wearing of breathing apparatus is encouraged. The indirect method of fire fighting does not lead to a free-burning blaze which burns off or vents such combustion products.

How to get information

Included with the SP program is a 10-page handout titled, "How to Obtain Emergency Response Information on the Southern Pacific Railroad—A Guide for Firemen." This describes the data available form the following sources:

A. On board the train itself. Besides waybills giving car contents, hazard classes, Standard Transportation Commodity Code, or STCC, numbers, placarding, etc., each SP freight caboose carries the 700-page book, "Emergency Handling of Hazardous Materials in Surface Transportation," issued by the AAR Bureau of Explosives. Each of 1675 different materials is listed alphabetically, with the basic nature of its hazard, fire attack for combustibles, and personnel protection needed. (The SP also distributed several thousand copies of this AAR book to all its on-line fire departments).

Each caboose also carries a poster explaining the DOT placarding system and the risks involved with each type of placarded cargo.

B. Railroad computer. Fire department dispatchers can contact SP officials by using a listing of eight 24-hour phone numbers, depending on the re-

Videotape is being made for Southern Pacific TV training program first used by the Portland, Ore., Bureau of Fire. Material videotaped was previously in a slide and tape cassette program. This scene is being shot on a car dome in a railroad yard.

gion involved, giving only the identifying initials and number of the freight car concerned, to request waybill data. If the car itself cannot be identified in a wreck, the locomotive or caboose numbers can identify the train. Numbers of the cars immediately adjacent to a derailed portion of train helps pinpoint the wrecked cars themselves.

The SP's computer system will then print out the complete hazard data from the AAR book.

Program praised

In a January 1979 letter to Southern Pacific, Chief Clyde Bragdon of the Los Angeles County Fire Department praised all this as "an outstanding multimedia presentation of proper emergency procedures for railway incidents...SP's training program and your company's attitude of preparedness will greatly improve our ability to work together in the control of any incidents on Southern Pacific rails with Los Angeles County."

"Very impressive," was the comment by Kern County, Calif., Chief Phil Anderson.

Material from this and the dome mobile training packages has been given to both the AAR and the U.S. Department of Transportation for use elsewhere in the country. In addition, fire science education courses at colleges in California, the Arizona State Fire School, and the California Division of Forestry's Training Academy have used the SP programs. Dome mobile training has been regularly presented at the Texas A & M Firemen Training School.

Altogether, contended Andre, "This has become one of the largest training efforts ever undertaken by a private transportation company."

Moreover, because of personnel turnover, Andre added, "Our training efforts are on an ongoing basis. Periodically, we return to each department to train new firemen and refresh those who have previously participated."

Vans ready with equipment

Recognizing the need for special tools, as well as training, to handle major emergencies, the SP has set up a fleet of eleven vans and trailers in locations from Houston to Roseville that contain hundreds of items of equipment. Because the Texas Gulf Coast area is a center of petrochemical manufacturing and transportation, several of these vehicles are kept in that region.

For example, a 40-foot trailer at Houston carries $54,000 worth of gear, including 10 all-chemical suits, lighting plants, 10 acid hoods, a four-man breathing air system, resuscitator, foam maker with 250 gallons of concentrate, power saws and other heavy rescue tools, 700 feet of hose, transfer and fire pumps, aluminized fire suits and coats, plus many pipe and valve tools. Three 15-foot trailers carry four-man breathing apparatus. Five ¾-ton vans carry explosimeters, acid suits, sets of chemical reference books, emergency gasket kits, extinguishers, turnout gear, breathing apparatus, and a variety of hand tools.

Much of this equipment can be sent quickly by air to remote locations where it may be needed. Backing it up are the experts of Andre's group.

One recent instance of this occurred in March 1979 when a 116-car train derailed at Lewisville, Ark. Among the ditched cars were six tankers filled with vinyl chloride and butadiene, one of which caught fire. Railroad personnel recommended evacuation of 1700 townspeople, while two members of the SP's hazardous materials control team in Houston were notified. They obtained information on car contents from SP's computer, advised Lewisville fire officials on tactics, then flew there by chartered plane to assist in directing fire fighting operations. Other SP experts were flown in from more distant points during the several days needed to clean up the wreck. No one was seriously hurt.

When such an emergency occurs along a main line in or near a town, access to the scene is not usually a major problem. In a large rail yard or terminal, however, the situation may be quite different (see Fire Engineering, August 1979, page 112).

Yard blueprints available

"We do discuss the question of yard accessibility," Andre said. "We urge that the local departments contact our engineering department and request a blueprint of the yard which includes fire hydrant and roadway or crossover locations.

"Most yards have designated an 'emergency spot area' which has water facilities and is relatively isolated from homes and businesses while still accessible by vehicle and large apparatus. These locations are where a leaking tank car, for example, or a boxcar on fire would be spotted. We also allow and support drills to familiarize fire fighters with yard layouts."

During 1979, the railroad videotaped a special combined version of the two audiovisual programs for broadcast via closed circuit TV to all 26 fire stations in Portland, Ore.

Added Andre, "We shot a new segment inside a caboose to show where waybills, train consists and emergency response information are found on a train."

This three-hour package will then be turned over to the Oregon state fire marshal for statewide distribution.

"Unfortunately," Andre pointed out, "we cannot take the dome mobile program off Southern Pacific lines. However, we do make the audio-visual programs available to all fire departments at cost, regardless of their location. I can be contacted for details on acquisition of the slide-cassette material."

His address is: C.R. Andre, Supt., Hazardous Materials Control, Southern Pacific Company, One Market Plaza, San Francisco, Calif. 94105. His phone number is 415-362-1212. ▫ ▫

Hose streams douse burning material peeled from dump by bulldozer. Chlorine gas evolving from burning PVC wastes complicated the extinguishment procedure—*photos by Dave Fornell.*

Dump Fire Complicated by Generation Of Chlorine Gas, Hydrochloric Acid

BY MEL GOODWIN

Most dump fires are merely nuisances, but one in Arkansas posed the immediate danger of chlorine gas generation and the added threat of hydrochloric acid in the runoff of water that might be applied to the fire.

The Gassville Volunteer Fire Department received a call on Tuesday night, Nov. 6, 1979, that there was a dump on fire about 8 miles north of town. (Gassville is 10 miles west of Mountain Home, Ark., near the Missouri border.) The fire department responded and found a private dump with a fire covering about an acre. The department expended its booster tank on the fire with no effect.

The local representative of the Arkansas Department of Pollution Control and Ecology arrived on the scene and said the dump consisted primarily of polyvinyl chloride (PVC) wastes and putting water on it would create hydrochloric acid runoff that might pollute the local creek and then the White River. The State Department of Pollution Control and Ecology has responsibility for dumps.

State agencies notified

The Gassville Fire Department then returned to its station. Both the Office of Emergency Services (state civil defense) and the State Department of Health were notified of the fire. The two agencies responded to the scene and, along with Pollution Control and Ecology, monitored the air and smoke in the area. In the area of the fire, some toxic fumes were found but not in any quantity. The air and smoke one quarter mile from the fire was free of toxic fumes. From Tuesday night to Friday morning, the wind was out of the southwest, blowing away from the populated area. On Friday, the wind shifted to the north, and blew the smoke over populated areas to the south and east. At this time, the health monitors did not detect any dangerous toxic levels more than 100 yards from the fire. By this time, the fire had involved about 2½ acres.

Inquiries led to the discovery that the owner of the dump had died on the Sunday preceding the fire. The materials in the dump were from Baxter-Travenol Labs, Inc. Baxter manufactures medical supplies, such as IV bags, tubing and needles. The waste polyvinyl chloride, along with goods that did not pass quality control, were disposed of under contract with the dump owner.

The dump was in hilly terrain on sandy, rocky soil. A ravine about 65 feet deep had been filled in with the waste, but apparently it had not been covered at regular intervals with dirt as required.

Problem discussed

The following Tuesday, Nov. 13, while I was director of the Arkansas Fire Academy in Camden, I received a telephone call from the Office of Emergency telling me an airplane would pick me up at noon to fly to the Mountain Home Airport with the OES director and the director of Pollution Control and Ecology. The three of us arrived at about 2:30 p.m. and surveyed the scene from the air and then on the ground. We discussed possible solutions, such as burying the dump (ruled out because the chlorine gas being generated was an oxidizer) or allowing the dump to burn out (ruled out because of the length of time involved and the steadily increasing toxicity of the smoke). When asked for my recommendation, I said the fire should be extinguished.

The problems associated with fighting the fire were discussed on the flight back to Little Rock. The distance to the river (1.8 miles) and the 500-foot rise to the dump ruled out the possibility of relay pumping. Several ponds were closer, so the possibility of using these was discussed.

Wednesday morning, a temperature inversion coupled with a rise in toxicity forced the evacuation of several homes in the area. With this development, Governor William Clinton declared a disaster and ordered state agencies to take necessary actions to eliminate the hazard.

The Fire Academy was directed to proceed to the dump fire, approximately 260 miles from Camden, with equipment to extinguish the fire. As director of the Fire Academy, I was placed in charge to organize the area volunteer fire departments to combat the blaze. The National Guard sent a truck to Camden to pick up 3000 feet of 3-inch hose, deluge sets and other equipment. On the way north, the truck stopped in Little Rock at a fire equipment dealer and picked up 850 feet of 5-inch hose and the necessary adapters.

The Fire Academy responded with a 1000-gpm quint, a van and the director's station wagon.

Strategy meeting held

Two academy instructors and I arrived with the equipment about 6 p.m. Wednesday and we set up a strategy meeting with the local fire chiefs. At this meeting, it was decided not to relay pump because the ponds were low from lack of rain. The fire departments were polled for equipment and it was discovered that one had a portable folding tank. It was decided to truck the water to the site and use the academy's quint to draft from the portable tank.

It was decided to use the academy equipment so as not to tie up the local fire department's equipment. It was also decided to use dozers to separate the burning material and to wet it down with 1½-inch hose, using fog patterns. This would limit the amount of water used and also prevent excessive runoff.

After agreement by the various agencies, the operation was started by 10 p.m. Wednesday. The Office of Emergency Services provided coordination and communications via repeaters. The Department of Pollution Control and Ecology monitored runoff, constructed dams and neutralized the runoff. The Department of Health monitored the air quality. The Arkansas Highway Department provided two dozers and dump truck tankers to haul water. The local fire departments provided manpower. These men were paid by the State of Arkansas for their time. The Arkansas National Guard provided the emergency lighting. The Arkansas Fire Academy supervised the fire fighting.

Hot spot is bared by bulldozer for extinguishment by careful use of hose stream to keep runoff water containing hydrochloric acid to a minimum.

The dozers worked from the upwind side and peeled the burning material from the pile as fire fighters wet down the material in front of the blade. A second crew kept a water curtain between the fire and the dozer. Two teams operated in this manner until the fire was extinguished about midnight, Friday, Nov. 16.

More tankers obtained

It was discovered early Thursday morning, when the operation was well under way, that the tankers could not keep up the water supply, so a local gasoline distributor was contacted to supply three 8000-gallon semi-tankers to haul water. This solved the water supply problem and when the operation was completed, a total of 465,000 gallons of water had been hauled.

The dump was very slick because of the water and the plastics. Another problem was the surgical needles that had been dumped in large quantities. The dump was also in a ravine on a steep hillside and this added to the complications. No one was injured, although some hose was punctured and when tested later, it looked like "lawn-soaker hose." Also, many pairs of boots were ruined. This pointed out the importance of proper protective clothing, such as steel insoles.

The deepest burn in the dump was only 6 feet after burning seven days. Since the dump was up to 65 feet deep, it would have burned for a long time. The toxicity greatly increased by Friday and early that morning, a temperature inversion made eyes water in Mountain Home, 10 miles away, and forced fire fighters to leave the dump for over an hour.

Runoff neutralized

The runoff was contained in two holding ponds and was neutralized with soda ash. The runoff had a pH of 1, which is a strong hydrochloric acid. As of last March, no water wells have shown any contamination.

The entire operation went smoothly because of the good cooperation among the agencies. Some local volunteer fire fighters took off from their jobs to provide the necessary manpower for the hard and dirty work. The Cotter Fire Department provided its new 1000-gpm pumper to relay pump as lines were extended to the rear of the fire. In all, men and equipment from 11 fire departments worked on the fire in shifts. The operation was continuous from Wednesday evening until midnight Friday, except for the hour Friday morning. Work resumed at 8 a.m. Saturday to check for hot spots and the hard work of picking up.

The State of Arkansas paid for equipment rental, including the Cotter pumper and the gasoline tankers. It also paid for all expenses of the state agencies and an hourly rate to the volunteer fire fighters. The total bill came to approximately $28,600. ☐ ☐

240,000 Forced to Flee Chlorine Released in Canadian Rail Wreck

BY GARY WIGNAL
and
JOHN KARL LEE

A set of wheels that tore loose from a car in a 106-car Canadian Pacific Railway (CPR) freight train set in motion North America's largest peacetime emergency operation.

CPR train 54 originated in the Canadian city of Windsor, Ontario, and at Chatham, 51 cars were added, including propane and chlorine tank cars. About 2 miles from the derailment site in Mississauga, just west of Toronto, an axle bearing overheated. The axle seized up and broke, hurling a set of wheels into a yard along the tracks. As the eastbound train neared Mavis Road, 25 cars derailed. Nineteen of these cars contained hazardous chemicals, chlorine, propane, caustic soda, toluene and styrene.

The first of five explosions occurred at 11:53 p.m. last Nov. 10, a Saturday, and the light from the explosion was seen at Mississauga Fire Department Headquarters, Station 1 at Highway No. 10 and Fairview Road. A deafening roar was heard and shock wave ripples were felt at the station.

Disaster plan activated

The fire fighters immediately put into action the disaster plan that Mississauga Fire Chief Gordon Bentley had recently formulated.

Just before fire fighters arrived, another explosion sent flames and about 70 of the 90 tons of chlorine in a tank car high in the air. This blast was so powerful that one half of the propane tank car rocketed 2300 feet through trees and buried itself 20 feet into the ground. The shock waves traveled along the ground for over 2200 feet and destroyed buildings, trucks and telephone poles.

As fire fighters began stretching lines, several more explosions occurred that sent the fire fighters for cover and knocked some of them off their feet.

Deluge sets used

The job of controlling the flames and cooling the propane and chlorine tank cars that were leaking began by first using 2½-inch hand lines and then moving to master streams from 12 deluge guns set up around the perimeter of the wreck.

At 1:30 a.m. Sunday, the first evacuation order was given to move 6000 people out of their homes on the west and south sides of the site. This was accomplished by police and ambulance personnel driving through the affected area and warning people through the vehicle public address systems.

At the same time, the Provincial Ambulance Coordination Center ordered all off-duty ambulance personnel in the Toronto area back to work. Region of Peel Police Chief Doug Borrows ordered all off-duty policemen to return to work and also asked for assistance from Metro Toronto, Ontario Provincial and Royal Canadian Mounted Police to help evacuate and secure the area.

Evacuation affects 240,000

During the next two days, the evacuation area was expanded seven times to include more than 240,000 residents and an area of over 20 square miles. "The train," as they were called over the next week, were moved to hospitals, nursing homes, shopping centers, schools, churches, service clubs and private homes that volunteered space.

The Mississauga Fire Department

Toronto Star Syndicate photo
Propane burns furiously after railroad wreck in Mississauga, a city bordering Toronto.

Wide World Photos
Deluge gun applies stream to cool derailed tank car, one of several in wreckage.

committed most of its companies to the derailment, and other fire departments offered assistance. The Etobicoke Fire Department, on the east side of Mississauga, sent companies to fill in at some of the vacant fire stations and Etobicoke Chief Bryon Mitchell responded to Mississauga Fire Headquarters to help coordinate communications.

Toronto International Airport sent a crash truck, but it was not used because of the intensity of the flames and the decision to control the fire rather than put it out. Other fire departments, such as Toronto and Oakville, sent spare breathing apparatus and bottles, and the Brampton Fire Department sent its fuel truck. The Oakville Fire Department answered alarms in the west end of Mississauga.

In response to Bentley's request, a Chlorine Emergency Plan team responded. This is a group of experts from the chlorine industry that responds to any incident anywhere in Canada involving chlorine.

Patch attempted

By Monday, the propane flames were low enough to allow workers to put a steel patch on the chlorine tanker so the remaining chlorine could be off-loaded to tank trucks. However, the heat and stress caused the railroad tank car sides to buckle so much that a good seal could not be made.

The "think tank" of experts at the scene decided on Tuesday, November 13, that the danger had subsided enough to allow residents on the outer fringe of the evacuation area to return to their homes.

Workers continued on Wednesday to remove parts of the derailed tank and and box cars that did not interfere with the chlorine tanker. With the propane tanker fire out, the remaining propane was transferred to a tanker truck.

On Thursday, five days after the derailment, the chlorine danger continued. Eight fire fighters checking hose lines and deluge guns stumbled on a pocket of chlorine and were taken to a hospital. All were released in a few days in good condition.

Air cushions hold patch

Later that day, a patch was devised by using inflatable air cushions between the chlorine tanker and the steel patch, which was held down by pieces of wood. Then the entire car was wrapped with chains to secure the patch. The air cushions were inflated successfully and a start was made to off-load the remaining 20 tons of chlorine into tank trucks, which contained neutralizing caustic soda.

At daybreak Friday, about 14 tons of chlorine had been removed and just after 3 p.m., 90,000 of the remaining 120,000 residents were allowed to return to their homes. But because 6 tons of chlorine were still in the tank car, the other residents had to stay clear of the area. At 7:30 p.m. the remaining evacuees were permitted to go home.

The command post was on Mavis Road, a quarter of a mile north of the derailment site.

On Wednesday, Nov. 21, the last fire apparatus left the scene after more than 250 hours of continued service by the Mississauga Fire Department.

There were no deaths and only minor injuries in the 11-day operation that included the evacuation of more than 250,000 persons from the area. ☐ ☐

Phosphorus trichloride fumes rise from accident scene in Somerville, Mass. At right is collecting pit for the toxic corrosive leaking from tank car against locomotive.

Boston Globe photo by Bob Dean

Toxic Spill From Tank Car Causes Evacuation of Mile-Square Area

BY ROBERT J. WILKER
Deputy Chief's Aide
Somerville, Mass., Fire Department

A railroad switching yard minor collision resulted in the release of highly toxic and corrosive phosphorous trichloride and required the evacuation of a mile-square area in Somerville, Mass. With a population of 80,596 in only 4.9 square miles, the city has an average population density of 16,448 persons per square mile.

At 9:10 a.m. April 3, the fire alarm communications office received two calls from the Boston and Maine Railroad dispatch office, reporting that a diesel locomotive had crashed broadside into a tank car at the B&M switching yard at Joy and Washington Sts. A large toxic cloud was overhanging the area and B&M workers were fleeing.

The fire alarm office dispatched three engines, two ladder trucks and Deputy Chief John P. Brosnahan and also notified the local ambulance service to respond with two units. Engine 3 and Ladder 1, located at Union Square, only four blocks from the incident, reported off at 9:12 along with Brosnahan.

Fire alarm radioed that the B&M reported that the tank car contained about 13,500 gallons of phosphorus trichloride, a corrosive, toxic liquid. While this information was being received, members of Engine 1 and Ladder 1 were removing a B&M worker trapped by the toxic cloud.

Use of masks ordered

As more information about the chemical was being obtained from CHEMTREC by the fire alarm office, Brosnahan ordered all companies to wear full protective gear with masks. Companies were ordered to lay 2½-inch hand lines, but not to charge them at this time. CHEMTREC reported that the chemical was water-reactive and that large amounts of water could be applied to the runoff—but not to the tank car.

Upon learning this, Brosnahan ordered a hole dug by B&M workers with front loaders to catch the spill running down the hill. B&M workers were given a crash course in the use of breathing equipment. However, not all of the spill flowed into the hole. Some was running down the hill to Joy St. and the sewer. The phosphorus trichloride was reacting with the damp ground and creating huge white toxic clouds in the area.

District Chief Willis Green, who responded to the Joy St. side of the spill with an engine and a ladder company, reported to Brosnahan by portable radio that the chemical was running down Joy St. toward the sewer. At this point, it was decided to open up the hand lines to dilute the runoff.

Evacuation necessary

Realizing that the water would intensify the toxic cloud, Brosnahan ordered the evacuation of an area of over 1 square mile, which included a Holiday Inn, a shopping center and several schools.

An engine and a ladder company were dispatched to the Holiday Inn to assist in the evacuation while police with loudspeakers evacuated the schools and homes.

After close examination of the tank car, Brosnahan knew there would be no chance of putting on any kind of patch to stop or even slow down the leak. The gash in the tanker was 3 feet high and 2 feet wide. After consulting with B&M officials and Chief of Department Charles Donovan, it was decided to bring in an empty tank truck to pump off the remaining chemical in the tank car.

Fumes shift with wind

Meanwhile the clouds of toxic fumes were shifting with the wind. The cities of Boston and Cambridge were notified of the potential danger heading their way.

As the morning drew on, fire fighters, along with scores of civilians, were taken to hospitals for treatment of skin irritations and inhalation of the toxic gas. The most seriously affected were the members of Engine 1 and myself. We had spent nearly all morning digging a trench to channel the phosphorus trichloride from the tank car leak to the hole being dug by front-end loaders. At various times throughout the morning, we were completely involved in the toxic cloud while digging the trench with hand tools. Every man in this operation exhausted over four tanks of air while digging. At 11:50 a.m., we were all taken to Somerville City Hospital by ambulance. The others were treated and released, but I was held 24 hours for treatment of first and second degree

Deck pipe is used by Somerville Wagon 2 in attempt to dissolve toxic cloud.

Photos by Ed Fowler

Chemical firm employee observes tank truck drafting phosphorus trichloride from pit.

Locomotive remains at switch where it punctured tank car and caused chemical spill.

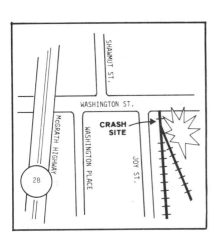

burns of both thighs, apparently caused by contact of the chemical and perspiration. I was on injury leave for 29 days.

Chemical neutralized

As the day went on, various EPA, state and federal officials arrived. How to remove the more than 6000 gallons which were now in the hole was discussed. The civilians proposed a massive flooding with water. Brosnahan knew the effect this would have and argued against it. The deputy chief wanted as much of the liquid as possible pumped out into a tank truck and the balance covered with sand and soda ash. At first he was opposed by several EPA, state, federal and B&M officials, who had now ordered the evacuation of almost half the city.

After careful consideration, it was realized that Brosnahan's proposal to use sand and ash was in fact the best idea and that operation was carried on throughout the night.

After he knew things were under control, Brosnahan, red-eyed and weary, went to the hospital for treatment. His expertise was a major factor in the control of the chemical spill.

All of Somerville's seven engine and four ladder companies were at the scene throughout the day. All members who worked at the scene were given blood tests and chest X rays at Somerville City Hospital. A total of 45 officers and men were examined.

Two men still out

At the time this article was written, it was questionable whether two fire fighters will be able to return to duty because of damage to their lungs. Several others were being treated for liver problems and inhalation of phosphorous trichloride.

During this operation, 17 cities and towns sent self-contained breathing apparatus and cascade systems to Somerville. Along with Commissioner George Paul, the Boston Fire Department sent Rescue 1, a district chief, and the department's chemist. All this assistance was handled by Metro-Fire, the 26-community mutual aid organization headquartered in Newton, 10 miles west of Somerville.

First-alarm apparatus was damaged by the fumes while directing fog streams from deck guns to try and control the vapor cloud spread. The wind shifted numerous times and apparatus was enveloped in the toxic cloud. The hose wagon of Engine 1 was placed out of service because the diesel engine was pitted by contact with the chemical fumes, and the Hurst Tool on Ladder 1 also had to be taken out of service because it also was pitted. The fumes also turned the lime yellow paint white on exposed apparatus. ☐ ☐

Computer Put In Vehicle of Haz-Mat Unit

BY BATT. CHIEF JAMES M. BRADY
Hillsborough Co., Fla., Fire Dept.

A computer portable enough to fit on the apparatus seat is carried by our hazardous materials response team. On the scene anywhere, it can provide ready reference to information on thousands of materials.

The Hillsborough County, Fla., Fire Department established the special response team in 1979 and began to equip a vehicle with useful equipment. One problem was storing and handling the many necessary books with technical information on hazardous materials. A common solution—or rather compromise—has been to select a limited number of the better publications and put them somewhere on the apparatus as a mobile library.

There are some disadvantages with this solution. For example, some good information had to be left behind after any selection. With a particular chemical, the text from a number of the books would have to be looked up and compared or combined for a full description. Handling the books and looking through indexes and thousands of pages would take too long. Pages would soil and tear, and the visibility of small print at night would be difficult.

We began to look at low-cost microcomputers for compact storage, rapid sorting and an easily readable display for the selected information. This had to be the answer for us, but it was not so simple as originally perceived because there was much to learn about microcomputers.

Before we could proceed we had to answer these questions: Should we use a split system with the main computer at the communications center and a keyboard/printer on the apparatus with radio access? Or should we use a self-contained on-board computer system, which would require the team members to be more actively involved?

Since cost was a great factor, we chose to experiment with the self-contained microcomputer. In selecting the type of on-board system to use, three basic criteria were considered. The system had to be:
- Relatively inexpensive
- Small in size
- Simple to operate

Our system cost less than $2000, including mobile installation. It is a TRS-80 computer from Radio Shack, with 32K memory (very roughly, 32,000 characters). Other models, with less memory capacity and cassette tape memory storage, were available at lower cost—but we wanted the flexibility and access speed provided by our disk memory storage.

Our microcomputer system did present some minor problems in converting to mobile application (such as providing the proper electrical line feed), but they were quickly overcome.

Software solution

Programming was the next item for selection. There are many software packages available that were already programmed for inventory control, accounting and other file-and-access uses. We could find no ready-made software package, however, designed to meet our special needs. Instead of adapting an existing program to our operation, we chose to create our own.

Not being programmers, we sought assistance from outside agencies who use computer systems on a large scale. With the help of a few interested computer enthusiasts, we were able to produce a suitable program for indexing, storing and retrieving hazardous materials information—in seconds.

One disk—several books

The contents of several books fits on a single 5¼-inch-diameter disk that looks like a flexible phonograph record. To load more information into the available memory, we developed a special indexing system to use our own terminology codes and abbreviations. We are in the process of rewriting a more sophisticated program to allow cross referencing and cross indexing.

We are also experimenting with a mobile telephone interface which will allow our microcomputer to communicate nationwide with other computers' information banks. The microcomputer is already to do many more useful things as we learn more about it.

Pre-fire planning is the answer to identification of gasohol and unleaded gasoline tanks during an emergency at a petroleum products tank farm.

Photos by John E. Bowen

Gasohol pumps, increasing in number, present new problems to fire fighters.

On-Site Gasohol Blending Breeds Widespread Extinguishing Problems

BY JOHN E. BOWEN

Most people know that gasohol is a blend of gasoline and alcohol but, beyond that, what is this new fuel? What problems will it generate for the fire service? Can we handle these problems with today's techniques and equipment or are new fire fighting methods going to have to be developed quickly?

Let's tackle these questions one at a time. First, what is gasohol? Gasohol is a mixture of unleaded gasoline and an alcohol, usually but not always ethyl alcohol, or ethanol, as it is sometimes termed. Methyl alcohol (methanol) or isopropyl alcohol (isopropanol) can be substituted for the ethanol.

So, before we can define gasohol precisely, we must define unleaded gasoline. The octane rating of gasoline, a measure of its knocking properties, must be increased for it to be used in today's autos. For decades, tetraethyl lead (TEL), and occasionally lead or tetramethyl lead, was added to gasoline for this purpose. Lead is a deadly poison, though, and a potentially serious environmental pollutant. Also, lead is destructive to the platinum catalyst used in emission control devices on late model vehicles. Thus, these late-model vehicles are designed to accept only fuel containing no lead—gasoline from which TEL and related compounds have been omitted.

The requirement for high octane gasoline remains, however, so a search was conducted for other chemicals that would raise the octane rating just as the discontinued lead compounds did. Tertiary butyl alcohol (TBA) was found to serve this function well and today TBA is the most commonly used anti-knock chemical additive. It is being added to gasoline at concentrations of 7 to 10 percent.

Methyl tertiary butyl ether (MTBE) is also an effective anti-knock agent. It, too, is used at concentrations of 7 to 10 percent. There is presently some research being conducted on use of TBA and MTBE in combinations totaling 15 percent.

When you drive into a service station to purchase unleaded gasoline, you're actually buying a blend of gasoline and TBA, MTBE or both. This is what we know as unleaded gasoline. As an interesting aside, unleaded gasoline is also likely to contain significant amounts of some aromatic hydrocarbons such as xylene, toluene and cumene.

Witches' brew

Returning now to the definition of gasohol, we begin to see that it's really a witches' brew of unleaded gasoline, alcohol, TBA, MTBE, etc. The ethanol that is used is 198 proof (99 percent pure) and would be prized for human consumption so the United States government mandates that it be denatured (made unfit for human ingestion) by addition of 4 percent methylisobutyl ketone. So, yet another chemical is added to the brew.

The gasohol being marketed presently in Texas, Louisiana, Hawaii and several midwestern states contains 10 percent denatured ethanol—one part alcohol to nine parts of unleaded gasoline. There is nothing particularly sacred about using 10 percent alcohol. It could just as well be 15 or even 20 percent. There would still be no major engine adjustments necessary. By adjusting the carburetor and compression ratios, though, pure ethanol could be used. In fact, this is being done on a small scale in Brazil and the midwestern United States.

From the fire fighter's viewpoint, unleaded gasoline and gasohol present problems that have seldom occurred before. Previously unknown chemicals will be brought into your first-due district. You likely know little about their fire potentials, vapor densities, lower and upper explosive limits, etc. Furthermore, do you have any idea how to extinguish fires in spills or tanks of TBA, ethanol, methylisobutyl ketone and so on?

Perhaps you're thinking that these chemicals are apt to be used mainly in refineries and bulk storage tank farms. Don't be too secure in this assumption. There are eight manufacturers of ethanol in the United States and 16 plants make TBA. Fifty-five more plants produce other gasoline additives—corrosion inhibitors and anti-freeze agents. These materials must be transported to the blending site by rail, truck and ship. The final products must then be moved to the neighborhood service stations.

Gasohol blended locally

Furthermore, gasohol cannot be stored for long because it absorbs water readily. When the critical water tolerance (the maximum amount of water permissible in the blended fuel) is exceeded, phase separation occurs. The alcohol-water layer has a higher specific gravity than gasoline, so the former sinks to the bottom of the tank and the gasoline floats on it. When this happens, we no longer have gasohol.

Gasohol, therefore, must be blended as close to the consumer as possible. Local distribution centers are now going to have a tank of ethanol on hand. There are going to be pipes running from this tank to the loading rack. Actual blending will likely be done right in the tank truck. This necessity for on-site blending will bring alcohol storage to many towns and villages across the nation. With ethanol storage comes a new problem for fire fighters—how do you extinguish it?

The same question holds true for TBA, methylisobutyl ketone, MTBE and each of the other chemicals that we've mentioned. How do you attack and extinguish a spill or tank fire in one of them? How can you prevent a spill from igniting?

Additive properties

Let's consider first the properties of the pure additives that will concern fire fighters (table I). Four of these chemicals are alcohols, namely methanol, ethanol, isopropanol and TBA. Being alcohols, these are polar liquids, a very significant fact as we'll see shortly. These alcohols have several other properties in common also. Each is a highly flammable liquid at normal atmospheric temperature. Although these alcohols are not subject to spontaneous heating, each will ignite when contacted by oxidizing materials. The vapors of each are heavier than air and pose a moderate explosive threat. Fortunately, none of these materials is dangerously toxic.

Methylisobutyl ketone is not an alcohol, but it is a flammable polar solvent liquid nonetheless. Its properties are much the same as those of the alcohols, though (table I).

Low-molecular-weight ethers, such as MTBE, are polar substances, too, but less so than the alcohols. Thus, they should in theory be less destructive to

Table I. Properties of selected polar liquids that are of interest to fire fighters.

Polar Liquid	Molecular Weight	Flash Point, °F	Autoignition Temperature, °F	Percentage LEL	UEL	Vapor Density
Methanol	32.04	52	725	6.7	36.0	1.11
Ethanol	46.07	55	793	3.3	19.0	1.59
Isopropanol	60.09	53	—	2.0	12.0	2.07
Methylisobutyl ketone	100.2	73	858	1.4	7.5	3.45
Tertiary butyl alcohol	74.12	52	896	2.4	8.0	2.55

foams than other polar solvents. Ethers are extremely volatile, though, and the resultant high vapor pressure may make it difficult to effect a foam seal against hot metal surfaces. More about this later, however.

So, what do we do when summoned to an alcohol, ketone or ether fire? These are polar solvents, remember, so conventional protein, AFFF and fluoroprotein foams will not be effective. Instead, we must use one of the specialized alcohol-resistant, or polar solvent liquid, foams.

Two types of PSL foams

There are two general types of polar solvent liquid (PSL) foams. One is a modified protein foam, but this is being phased out slowly by the manufacturers. Let's look at the second type, the modified AFFF which we'll call PSL-AFFF. Several manufacturers produce these alcohol-resistant foams in the United States, but each of them is a modified AFFF. Thus, the designation PSL-AFFF refers to this general type of foam concentrate and not to a specific brand name. These are the only foam concentrates available today which will extinguish alcohol, ether and ketone fires.

PSL-AFFF concentrates contain a polymer which precipitates out of the foam blanket to form a layer over the fuel surface, thus effectively separating fuel vapors and oxygen. The PSL foams will extinguish hydrocarbon fires as well as polar liquid ones, but the technique of applying PSL-AFFF to hydrocarbons is somewhat different than that of conventional AFFF.

PSL-AFFF may be applied to burning hydrocarbons with either air-aspirated or non-aspirated nozzles. Further, it can be applied to the hydrocarbon via topside or subsurface methods. Consult the manufacturers' literature, though—the recommended application rate for a given PSL-AFFF may differ for PSL and hydrocarbon fires.

When a PSL-AFFF is used on a polar solvent fire, it must be air-aspirated. There is no choice about this if maximum effectiveness is to be achieved. Furthermore, PSL-AFFF must be applied topside to a polar solvent liquid fire. Even the alcohol-resistant foams are destroyed when immersed in the liquid. Thus, they cannot be applied via subsurface injection.

Swedish technique

Swedish fire fighters have developed a technique that permits injection of PSL-AFFF into the bottom of a fuel tank, a technique known as semi-subsurface injection. It hasn't received much attention in the United States yet, but generally here is how it works. The foam is pumped into a pipe on the bottom of the tank. The pressure ruptures a seal and pushes a plastic tube to the surface of the burning fuel. The tip of the tube burns off, gently releasing the foam to extinguish the fire.

PSL foams must be applied gently to the burning fuel. In the case of an alcohol spill fire, for example, direct the foam stream at the ground in front of the fire. Bounce the foam onto the fuel surface. For topside application to a tank fire, deflect the foam off the opposite side of the tank. Failure to apply the foam gently will cause the foam to be mixed with the PSL fuel, thus destroying the blanket before it can extinguish the fire.

Another problem is caused by the high viscosity of PSL-AFFF concentrates, a much greater viscosity than that of standard AFFF. The viscosity of the concentrate is such that standard in-line proportioners with ⅝-inch diameter pickup tubes may not be able to supply the concentrate fast enough to achieve a 3 percent application rate in a 60-gpm or greater stream. The foam that you actually get at the nozzle may be only 2 or 2.5 percent.

If your proportioner is continuously variable from 3 to 6 percent, set it at 3.5 or 4 percent. For those proportioners that cannot be varied, it may be necessary to replace the small diameter pickup tube with a larger one, e.g., 1¼-inch diameter, for use with PSL-AFFF. Do be aware of this problem and check the pickup tube frequently when using PSL foams. Different makes of PSL-AFFF may differ as much as 10-fold in their viscosities, so this problem may occur with one concentrate but not with another.

Apply alcohol-resistant foams to fires in pure PSL additives at rates of 0.15 to

0.20 gpm/ft² of surface area. Methyl, ethyl and isopropyl alcohols are among the most difficult of the polar liquids to extinguish and require a high application rate.

The radiated heat from a PSL fire will be much greater than that from an equal volume of burning hydrocarbon. This may magnify exposure problems as well as accelerate thermal destruction of the foam. And don't forget that there is no visible flame when methanol burns!

Fire in an MTBE bulk storage tank requires topside application of a PSL-AFFF even though MTBE is not as polar as the alcohols. Ethers are not as foam-destructive as alcohols, but even their limited destructiveness dictates use of a PSL-AFFF.

Ethers are much more volatile than alcohols, though. For example, the vapor pressure reaches 400 mm at 112°F, 146°F and 186°F for methanol, ethanol and isopropanol, respectively. The corresponding temperature for MTBE is 106°F.

MTBE vaporizes rapidly at fire temperatures, resulting in very high vapor pressures. The high vapor pressures make it difficult to maintain a foam seal at points where the liquid touches hot metal. The expanding vapor continuously disrupts the blanket. This problem can be readily eased by supplemental cooling of the metal surfaces with hose streams, a procedure that produces quick control of the fire.

PSL-AFFF is effective in securing either hydrocarbon or polar solvent spills before ignition occurs, forming excellent blankets with lengthy drain-off times. The foam blanket will also significantly reduce the potential toxic and environmental pollutant effects because evaporation is greatly decreased.

Let's briefly summarize the information presented thus far. Remember, the TBA and MTBE being added to gasoline to produce the unleaded product are polar liquids, as is the methylisobutyl keton used to denature ethanol for use in gasohol production. The alcohols themselves are also polar. Therefore, the only effective extinguishant for fires in these pure liquids is a polymeric polar solvent foam, a PSL-AFFF concentrate. No other foam will do the job! And, remember when using PSL-AFFF products that conditions of use and application methods differ from those for conventional AFFF.

Amount of foam needed

Let's give some thought to the final products now, the unleaded gasoline and the gasohol. Spill fires of unleaded gasoline containing up to 10 percent TBA or MTBE can be extinguished with any foam—protein, AFFF, PSL-AFFF or fluoroprotein, applied at 0.2 gpm/ft². At this rate, spills of 10, 25 and 50-foot diameters will require the application of 16, 100 and 400 gpm of foam, respectively. Don't underestimate this requirement when you make the initial attack.

PSL-AFFF is the extinguishant of choice for subsurface injection into tanks of burning unleaded gasoline. It must be air-aspirated and applied at a rate of 0.10 to 0.15 gpm/ft². Thus, 50 and 100-foot-diameter tanks require 200 to 300 gpm and 800 to 1200 gpm, respectively.

PSL-AFFF applied topside at 0.15 gpm/ft² is also very effective on tanks of burning unleaded gasoline. Fluoroprotein foams will do the job under these conditions, too, but the application rate must be increased by 33 percent.

Gasohol spill fires up to 1 inch deep can be extinguished with AFFF if the alcohol content of the blend is 20 percent or less. It is likely that more foam will be required than if PSL-AFFF were used, though, because of alcohol-induced AFFF destruction. PSL-AFFF must be used if the alcohol content of the gasohol blend exceeds 20 percent. It is suggested, however, that a PSL-AFFF be used on all gasohol fires regardless of the alcohol percentage. Apply the foam at 0.2 gpm/ft².

Gasohol tank fires must be attacked from the top, pending further progress in alcohol-resistant foam technology. PSL-AFFF and fluoroprotein foams will both extinguish the fire, but the latter must be applied at a higher rate—i.e., 0.15 gpm/ft² for PSL-AFFF vs. 0.2 gpm/ft² for fluoroprotein foam.

Securing gasohol spill

How can you secure a gasohol spill to prevent ignition? The PSL-AFFF foams will certainly work, as will the fluoroproteins. Protein foam and the AFFFs may do the job, but this depends on the specific brand of foam and also on the source of the unleaded gasoline. Some combinations give excellent results. With others large tears and blisters quickly appear in the foam blanket. Thus, plan to use a PSL-AFFF or fluoroprotein for handling gasohol spills.

It has been suggested in various articles recently that gasohol fires could be extinguished by initial phase separation followed by conventional (non-PSL) foam attack. We've mentioned that addition of water beyond the critical tolerance to gasohol will cause phase separation. Alcohol is almost infinitely soluble in water and the specific gravity of water being greater than that of gasoline (1.00 vs. 0.66 to 0.69), the water-alcohol phase will sink to the bottom of the storage tank or spill. "Pure" gasoline will float on the surface.

Theory has it that a sweep of the burning gasohol surface with fog streams will cause phase separation. You would then be confronted with a gasoline (not gasohol) fire that could be extinguished with AFFF, protein or fluoroprotein foam.

There are some rather obvious problems with this technique, though, and further study is definitely need. For instance, it certainly cannot be done to a full storage tank—there's no place for the water to go. If attempted on a gasohol spill fire, it would be all too easy to start the fire flowing. Lastly, the water absolutely must be stopped before foam application begins. Otherwise, the foam blanket would be disrupted quickly.

Phase separation with water is therefore not recommended as a fire-fighting technique for gasohol fires at this time. Our knowledge of potential undesirable effects is inadequate.

Carry flammable placards

Before we bring this discussion to a close, though, two more factors must be mentioned briefly. First, unleaded gasoline and gasohol shipments will be placarded as flammable liquids. No further identification of the product is likely to appear on the vehicle. The incident commander must obtain this information from the shipping papers, i.e., the shipping order, bill of lading, cargo manifest or waybill.

The product involved in the fire or spill must be specifically identified before any control measures can be taken. Thus, it is critically important that the commanding officer promptly obtain this information. Failure to identify the product correctly can negate all further actions.

Secondly, the fire attack procedures detailed in this article supplement basic tank and spill fire-fighting procedures. They do not replace the standard size-up and tactical procedures for fires in flammable liquid fuels.

Finally, know where polar liquids, unleaded gasoline and gasohol are stored, sold and used in your area. Pre-plan for these emergencies. Transportation incidents are another story. They cannot be pre-planned. Remember, though, to try to identify the product immediately.

THE AUTHOR: John E. Bowen received 10 years of fire fighting experience with the College Park, Md., Fire Department and the Laurel, Md., Rescue Squad. He has a Ph.D. in biochemistry from the University of Maryland and moved to Hawaii 15 years ago. For the last 10 years, he has been in the University of Hawaii fire science program. His special interests are in the area of hazardous materials, which allows him to combine his fire fighting experience with his formal education.

Toxic, Flammable Chemicals, Gases Breed Trouble in Electronics Plants

BY R. L. NAILEN
Staff Correspondent

What once filled a room now fits on the surface of a postage stamp. That sums up the progress of electronics since World War II. Computers are revolutionizing our civilization—including some fire service operations—because they can now be made so small that microcircuits and microprocessors are common terms in the industry.

Though its products are small, the electronics industry is huge and growing fast. Annual growth rates from 15 to 30 percent are forecast through 1983. Total value of chemicals used in producing integrated circuits and related components is approaching half a billion dollars a year and is expected to double by 1985.

The manufacture of such electronic components as semiconductors is no ordinary industrial operation. Typical processes are plasma enriched deposition, reactive ion beam etching, pulse electron annealing, and deep ultraviolet flash polymerization. These involve materials and methods which are toxic or explosive and can generate unpredictable combinations of deadly hazard. Recent events in Santa Clara County, Calif., have emphasized the importance of this to fire fighters everywhere.

Silicon Valley

The once agricultural flatland comprising the county's northern section is now known as Silicon Valley because of its massive concentration of semiconductor manufacturers (a billion-dollar investment) using silicon chips in their work. These chips are the basic building block of many electronic circuits—tiny bits of highly complex, precisely formulated chemical structure which perform the memory and calculation routines upon which data processing is based—from the pocket calculator or kid's electronic game to the giant computers that control a power plant, regional air traffic, or an entire oil refinery.

Scores of electronics firms of all sizes form the major industrial base of this area of half a dozen cities with nearly a million residents. Companies headquartered in Silicon Valley generate nearly half the entire country's output of integrated circuits, and 40 percent of the world's semiconductors. This is the ninth largest manufacturing center in the United States and the fastest growing one.

Santa Clara (pop. 100,000) contains many such companies. The buildings tend to be one-story, tilt-slab construction, often in parklike complexes having little conflagration potential and ample water supply. But besides the extreme danger of exotic chemicals and processes, they frequently include very large undivided areas with few access openings. Early in 1980, one computer component manufacturer occupied a building in Santa Clara containing 10 acres under a single roof with few windows or doors over much of its periphery. Plant security is a problem. Trade secrets—and the hardware itself—have been pirated many times in this highly competitive field.

Several fires yearly

How often does the fire service face problems at these facilities?

Fire Marshal Gary Smith of nearby Mountain View said, "We have had a fairly typical fire history in these plants. We have several fires a year . . . losses sometimes over $100,000. One of the most common causes . . . is immersion heaters igniting fiberglass dip tanks used on a plating line."

Bill Fleming, Santa Clara's fire marshal prior to his death last June, stressed the danger of chain reactions among processing chemicals following outbreak of a fire.

Said Fleming, "A major concern in most semiconductor fires is the mixture of chemicals and the unknown toxic results of their reactions when fire causes them to explode."

He added that, no matter how frequently buildings are inspected, "It is fairly common for a company to alter a building without taking out the proper permit. In those cases, things like sprinklers, fire extinguishers and good exit paths are usually overlooked."

Chemical spill problem

Actually the fire or explosion hazard is less than the chemical spill hazard. The gases and liquids, such as cyanide compounds, used for electronics manufacturing are unlike anything in the experience of most fire departments. These materials are used as plating solutions, doping gases, substrates and etchants. They include such toxic gases as arsine, phosphine, hydrogen chloride, and diborane, which have become of such concern to the semiconductor industry itself that a gases safety task force was formed in 1980 to prepare standards on their handling, storage and use.

Explosive hydrogen is sometimes involved. The pyrophoric gas, silane, is common. Silane bursts into flame immediately upon contact with air.

Hence, as Smith pointed out, "Once a leak is detected, it is too late."

According to a July 1980 survey of the type of research lab that abounds in and around Santa Clara, nearly half of all workers and two-thirds of the chemists regularly work with toxic and flammable compounds. Yet more than a third of all chemists surveyed reported that no overall safety program existed in their plants. Although most major firms do have a safety officer, the smaller companies do not.

Have to be prepared

The safety officer of one of Santa Clara's biggest electronics manufacturers commented, "Even with elaborate detection systems, the potential for fire still exists. You have to be prepared for an inadvertent spill of flammable liquids."

By the end of 1979, such dangerous spills were occurring in Santa Clara "almost weekly," according to Fire Chief Don Visconti. Two recent incidents there are good examples. The first took place April 15 at the Memorex plant and kept six fire companies busy for two hours. A valve malfunction permitted nearly 1000 gallons of highly flammable cyclohexanone solvent to flow into a diked yard. Exposures had to be covered, 300 employees evacuated, and hose lines manned until arrangements could be made to pump the liquid into a tank. One engine company checked a nearby residential area for fume odor, some of which was detected around an apartment building.

Second incident

The second, much longer incident occurred at the National Semiconductor Corp. late at night May 4. Nitric acid leaked from a 5000-gallon storage tank onto a copper pipeline containing compressed nitrogen. Escaping gas then dispersed a huge cloud of acid vapor into the air. No one could be found who knew where to shut off the nitrogen line, which was erroneously labeled "compressed air." Eventually, company officials were located who had a knowledge of the system, which included several other large nearby tanks containing sulfuric and hydrofluoric acids. Plastic shields around the diked tank area were broken away to get at the shutoff valves, after which chemical neutralization of the spilled material was completed. A pump, installed to transfer spills within the dike to a holding tank, was destroyed by the acid. One fire fighter suffered minor face burns and some equipment was contaminated.

There had been few fires in Santa Clara. However, within a six-month period, several multiple-alarm blazes occurred which led to drastic action by Visconti to deal with the problem of chemical hazards in general and electronics manufacturing specifically. The worst of these was a third alarm of electrical origin at International Materials Research last Jan. 22, which resulted in a million-dollar loss.

During the four-hour extinguishment of this fire by 10 companies of the Santa Clara Fire Department, toxic fumes of a still-unidentified nature, drifting in the large cloud of smoke, affected two police officers some distance away with facial rash and headache. Despite extensive use of breathing apparatus (55 air bottles) by the 70 fire fighters involved, three men were treated for nausea, headache, or rash. Four others were hospitalized with more conventional injuries.

"I was worried about the TV news helicopter hovering up near that smoke," added Fleming, "but apparently they didn't get too close."

Fumes spread to hall

During a recent two-alarm blaze in an electronics plating shop, fumes were drawn into the air conditioning system at a nearby meeting hall, where 15 occupants had to be treated for breathing problems. Said Visconti, "When I see large numbers of people requiring medical treatment like this, I am alarmed."

The Santa Clara chief found himself unable to rely on conventional building inspection methods or reports to keep abreast of hazards in this fast-growing industry.

Said Battalion Chief Bob Boeker, "They change their processes just as fast as you change your underwear."

Manpower limitations in Santa Clara have meant two or three-year intervals between inspections of some firms. In other Silicon Valley cities, inspections are no more often than annually.

Said another officer in one of those cities, "Our fire fighters can do a good maintenance inspection of a commercial building, but to inspect a semiconductor processing facility, you need to know the chemicals, the processes and the mechanical systems. And that requires a specialist."

Building survey made

So Visconti proposed a citywide chemical hazard assistance program. His first step, completed during May, was to take a complete chemical survey of every commercial building in Santa Clara—some 5000 in all. This required putting all his companies on the street six hours daily, six days a week, for a month. (Santa Clara operates 13 fire companies from eight stations with a paid force of 140, plus 50 active "volunteer reservists." Members of the latter group responded on all the incidents described here, 20 of them at the third alarm in January.)

"No other department in the nation has taken this approach," Visconti claimed, although several others in the county and the adjacent territory north toward San Francisco have recognized the problem and are trying to improve their communications with industry.

Here are samples of what the Santa Clara survey disclosed:

Inside storage of over 25 gallons of combustible liquids in 781 buildings; over 2000 cubic feet of flammable compressed gas in 525 buildings; poison gas in 48 buildings; toxic liquids in 233 buildings; over 55 gallons of corrosive liquid in 418 buildings; corrosive solids in 106 buildings; over 500 pounds of oxidizers in 123 buildings; and poisons or insecticides in 189 buildings.

Program objectives

Using the results of the survey, Visconti's program includes such steps as:

1. To create within the fire department a chemical hazard assistance unit, manned by specialists, to provide expertise within the department and to work with industry and the community for solutions to the chemical problems.

2. To create fire department operational plans for dealing with chemical hazard incidents.

3. To support the development of educational programs (working particularly with the Chamber of Commerce) for industry, fire department personnel and the community.

4. To support the creation of a "center of excellence" at Mission College, providing a West Coast facility for research, testing, and information retrieval relating to hazardous chemicals.

As part of a $162,000 budget for this total program, since accepted by the City Council, the Santa Clara chief asked for two chemical hazard assistance specialists and a chemical vehicle with trailer. Job specifications were being drafted in July, and a Chamber of Commerce task force was being put together as liaison between industry and the fire department.

This group, explained a chamber spokesman, "will address what types of hazards and toxic chemicals are out there, where they are located and how they are being watched over and protected now."

Visconti is optimistic that this kind of industry and fire service cooperation, combined with the new specialists under his command, will substantially raise the level of public protection in his part of Silicon Valley. ◻ ◻

Inspector Thomas Powell checks for use and storage of hazardous materials before a permit is issued—*Photo by the author.*

New Inspection Program Tracks Hazardous Materials

BY JOSEPH FERGUSON

Fire and building department officials in Poughkeepsie, N.Y., have received extensive training over the years in how to recognize and properly handle hazardous materials.

As the result of this training, a joint effort was launched by the fire and building inspection departments to establish a comprehensive system of inspections which would not only ensure proper use of dangerous substances by industries but would also record amounts and locations of all materials present.

Unfortunately for Poughkeepsie, this program was still in the planning stages when the explosion and fire of last Jan. 14 ripped through the Berncolors-Poughkeepsie Dye Works (see Fire Engineering, July 1982), killing two employees and precipitating the largest chemical spill of its kind in New York State history.

Top priority program

Upon completion of the month-long Coast Guard supervised cleanup operation, implementation of the inspection program became a top priority for Poughkeepsie Fire Inspector Thomas Powell and Building Inspector Michael Haydock.

In one way the Berncolors incident made the job of the inspectors easier since it served to shock the public into realizing that "It can happen here." "In fact," says Powell, "we received a unanimous response within three days of mailing our initial letter informing the local industries of our intent."

This letter, which was signed by Fire Chief James Davison, informed local businesses that a permit is required, according to the American Insurance Association (AIA) Fire Prevention Code of 1976, for the "storage and handling of certain materials which may be highly flammable, explosive, or contribute to other hazards." The letter explained that municipal records indicated such materials might be used in that particular business. Arrangements should be made immediately with the fire department, the businesses were told, so the matter could be reviewed.

Whose responsibility?

The mechanism and authority for implementation of this permit procedure is already written into Sections 20 and 21 of the AIA Code, which, according to Haydock, states clearly that, "It is not incumbent upon the city to search around for hazards but on the businesses using these materials."

Despite the burden for notification being on the property owner, a decision was made to identify all hazardous material sites within the city and to make inspections of those thought to be the most dangerous.

Powell and Haydock first went over all available lists of commercial properties including the rolls of the city assessor, records of the New York State Department of Environmental Conservation, and information collected by a local engineering firm contracted the previous year to study wastewater usage in the city. This study not only helped to locate industries in the city that might otherwise have been overlooked because of size or nature of business, but also gave some indication of the kinds of materials being used.

Permit application

At the same time, applications for hazardous chemicals storage and handling permits were drawn up according to the specifications of the AIA Code. Then the city's corporation counsel checked them.

These applications provide the inspectors with much information regarding the particular business in question: names of all owners or holders of mortgages or other liens, the nature of the business and types of materials handled (explosives, flammables, combustibles, oxidizers or corrosives).

No permit is issued unless manufacturers' data fact sheets are provided for all materials being stored or handled. These fact sheets are approved by the U.S. Department of Labor and are essentially similar to Form OSHA-20. They detail the hazardous ingredients of a particular substance as well as such physical characteristics as boiling point, specific gravity, vapor density, percent of volatiles by volume, freezing point, vapor pressure, solubility in water, evaporation rate, appearance and odor. They also list data on fire and explosion hazards such as extinguishing media, special fire fighting procedures, and unusual fire and explosion hazards.

In addition to this information, the permit applications must also include a scale drawing of the building and property. This illustrates the location of all materials being stored or handled, as well as the name of the person in charge of operations and his qualifications with regard to training, experience and knowledge of the materials being used.

The permit applications are hand-delivered by the fire inspector to each of the industries on the list. In this way he can explain in detail exactly what the code provisions require and how to go about filing the application.

All permit applications result in inspections, both to verify information received and to collect additional data important to the fire fighting operations, such as location of standpipes, fire hydrants and accessibility to apparatus.

Compiling the book

The information gathered through the applications, data sheets and scale drawings and inspections is then compiled into a comprehensive book which is to be carried on all city fire apparatus. This book is arranged alphabetically by street names and, where there is more than one industry on a street, numerically within the street. All information pertaining to a particular location has one place in the book. There is no confusing cross-referencing.

Permits are issued for a period of one year. Then the procedure must be repeated.

Thirty-two locations have been identified as users of hazardous materials. As new industries enter the city, they will be added to the list. When the "top priority" list is taken care of, less dangerous businesses such as gas stations and paint stores will also be given attention.

The continuous training of building and fire department personnel in the recognition and proper handling of hazardous substances will help to serve as a back-up to the system. Officials not normally connected with hazardous materials inspections—such as the plumbing inspector—will be able to recognize anything out of the ordinary. They may be better able to catch industries that were missed or that are trying to conceal certain activities.

Another plus of the system is that when violations are discovered, the court cases are relatively simple for the law department. Proper inspection was made, hazardous materials were discovered, records indicate no permit has been issued.

Generating publicity

Public awareness of the situation is still a prime factor. Poughkeepsie is trying to generate as much publicity about the program and the dangers of hazardous materials as possible. Publicity has a two-fold purpose, making people who do deal with hazardous materials more aware of what they are doing, and giving the general public the knowledge it needs to serve as a further watchdog over industry. The practicality of this reasoning has already proved its worth in Poughkeepsie.

Responding to a neighbor's complaint about a local metal-plating operation, Inspectors Powell and Haydock discovered a barrel of sodium cyanide being used as a garbage can on a street used by children as a shortcut.

Instructions on the label clearly stated that the substance could be fatal if swallowed or inhaled and that contact with eyes and skin should be avoided. In addition, it specifically prohibited the reuse of the container or the flushing out of its contents into the sewer system.

Assuming that contaminants were still in the container, Powell ordered a city sanitation crew, equipped with protective rubber gloves, to remove the barrel to a secure area in the public works compound to be held as evidence.

Meanwhile Haydock obtained a search warrant from Assistant Corporation Counsel Thomas Halley to inspect the rest of the premises.

An inspection conducted the next day in the presence of a company representative revealed a number of other violations including containers of poison and oxidizer improperly stored with trash in an unsecured shed, as well as improperly stored LPG tanks. They were found leaning against a fence in the yard and buried beneath thickly grown vegetation.

The discovery of these violations in a small business located in the heart of residential neighborhood serves to underscore the necessity of the new inspection program. ◻ ◻

The Process of Making Decisions At Hazardous Materials Incidents

BY CHIEF WARREN E. ISMAN
Director
Department of Fire And Rescue Services
Montgomery County

What happens when an alarm is received for an incident which involves the release or potential release of hazardous chemicals? Does the first-alarm incident commander decide to take the long way to the scene? Does he immediately submit a leave request so that he is relieved of potential decision-making responsibility? Well, these might seem like farfetched ideas, but really, how prepared is the first-arriving commander for handling an incident involving hazardous chemicals?

One of the basic problems is that the decision-making responsibility at a hazardous material incident is very different from handling a structural fire. While the basic goals of protection of life and preservation of property remain the same, the actual decision-making process is completely different.

The first-arriving commanding officer is experienced in handling structural fires because they occur frequently enough. But what about the hazardous material incident? The variety of chemicals and the potential locations of the incidents make them very different. The many varieties of chemicals and their combinations combined with the variations in locations and types of release make it almost impossible to develop experience and a cookbook approach to decision-making for the incident. Therefore, initial size-up and evaluation must be done in a systematic way so that the best response can be determined.

It is important for the incident commander to remember that at a hazardous material problem, doing nothing may in fact be the correct thing. The fire service has become involved in the incident because something has gone wrong, and that is certainly not the time to begin the study of the various methods and techniques for handling the incident. The initial arriving officer must be prepared in advance with the knowledge, information, and technique for mitigating the hazardous material incident.

Decision-making

Well, how does the initial incident commander organize his thoughts to begin a systematic analysis of the problem? The key here is the use of the phrase "systematic analysis." If this is done correctly, it should minimize the confusion which accompanies many large hazardous material incidents. In this way, the chief officer can prevent mistakes and injuries while taking corrective action.

Basically, the decision-making process involves the following steps which must be considered:

1. Is there a chemical hazard?
2. If there is a chemical hazard, what is the product?
3. What hazard does the identified product present?
4. What objectives have been established by the first-arriving chief officer?
5. Of the established objectives, what are the alternatives for accomplishing them?
6. Which alternative is the best?
7. Once the alternative has been implemented, is the decision correct or does it need to be modified?
8. When the incident is concluded, have personnel and equipment been decontaminated correctly?

The first step in the decision-making process for the initial arriving chief officer is to determine if, in fact, there is a hazard at the scene of the incident. Input for this decision comes from many sources, including:

1. Preplan information.
2. Placards on the sides of the transporting vehicle if it is a transportation incident.
3. A label on the outside of the shipping container.
4. An NFPA 704 symbol on the outside of the storage container.
5. The shape of the container, which may indicate a pressurized vessel or a specific type of chemical.
6. Observation of physical characteristics, such as the color of the storage container, an odor from the spill or leak, or a vapor cloud rising from the area of the incident.

Preplan information is obviously ex-

Systematic analysis of the problem should precede and guide the fire attack. At this multivehicle accident, fire fighters use 1¾-inch lines on the fire—*photo by Curt Hudson.*

tremely important. It is necessary for fire service personnel to get out into the community to determine the type and location of hazardous materials. In addition, special preplanning must be done for the various transportation routes through the jurisdiction. For highway transportation, highway access, terrain, water supply, and a general idea of the commodities transported need to be developed. The same information can be prepared for railroad transportation with the added problem that access to portions of the track may present certain difficulties. Fire officers must also remember that water transportation and pipeline transportation must be included in their preplanning efforts. Finally, transportation by air needs to be considered not only for airports in the immediate vicinity, but for an incident involving an aircraft while traveling through airspace over the city.

Weakness of labels

Placards and labels can provide general information about the hazard. The chief officer must remember that the rules governing the use of this type of warning system are complex. For example, on truck transportation for many commodities, it is necessary only to placard the truck when carrying over 1000 pounds of the hazardous material. Obviously then, the initial warning would not be present on the outside of the vehicle if only 900 pounds of a poison liquid were contained inside. It is also important for the officer to note that federal regulations require only one label on the outside of the shipping container. Thus, if the shipping containers are stacked one on top of another, the warning label may in fact not be visible.

There is a new placarding requirement for bulk transportation which involves the use of a four-digit United Nations number. The requirements and utilization of this kind of placarding are discussed further on.

The NFPA 704 symbol is a voluntary system for fixed storage devices. The system using numbers from zero (no hazard) to four (highest hazard) provides immediate information on the health, flammability and reactivity problems of the product stored. However, note that there is no way that this system can provide the specific name of the product.

The shape of the container may provide certain basic information and indication that there is a chemical hazard. Pressurized cylinders generally have rounded ends because of their ability to hold pressure. Relief valves can generally be seen on larger pressurized vessels. Some storage tanks, such as for cryogenics, have specialized shapes which can be easily identified. Other things, such as the shapes of specific

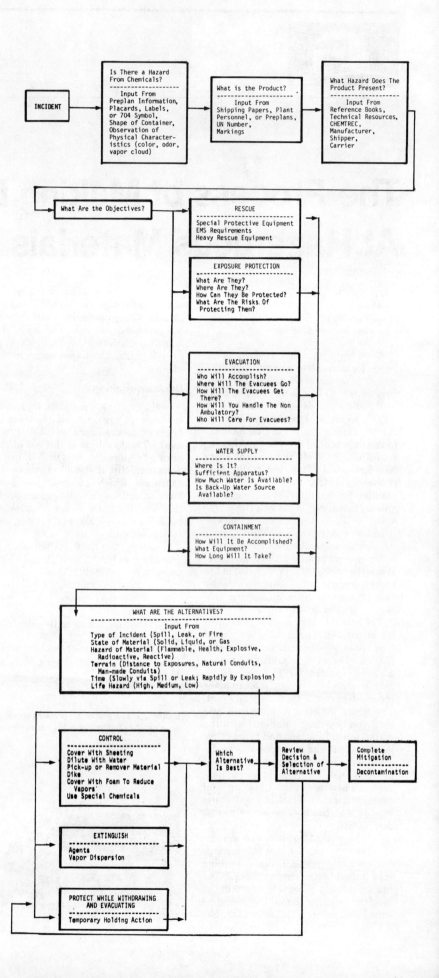

46

Full protective clothing allows fire fighters to intervene in more types of incidents. Here, a ruptured pipe released toxic fumes and forced the evacuation of thousands of workers. *Wide World Photos*

types of rail cars, can indicate that the container is for a hazardous chemical. It is important to remember that the shape of the container can give only a general idea about the potential problem and cannot be used as a specific guide to the product contained.

Finally, the initial chief officer can use his senses to determine a potential problem. The presence of an odor—particularly if it is irritating—or a vapor cloud, or a special color to the spill can indicate the presence of hazardous chemicals.

Remember, in this first step the chief officer is only making a determination that there is, in fact, a chemical release or potential release and that a hazard is present.

Product identification

Once the chief officer has determined that a hazardous chemical exists, the next step is to identify the product. It is necessary to know the specific chemical name if further information is to be obtained. Input for this decision is obtained as follows:

1. From the shipping papers if the incident involves transportation.
2. From plant personnel if fixed storage is involved.
3. From preplans developed by the fire department.
4. From placards if the United Nations number is shown.
5. From markings on the outside of the transporting vehicle or the storage container.

The transporting vehicle carries shipping papers which may indicate the specific chemical name of the product. In rail transportation, the shipping document—called a waybill—is carried by the conductor either in the caboose or the engine. In truck transportation, the shipping document—called a bill of lading—is kept by the driver in the cab. When the driver is absent from the truck, the shipping documents should be placed on the driver's seat. In air transportation, the air bill is kept by the pilot.

Manifest aboard ship

For water transportation, the master or mate of the ship carries the dangerous cargo manifest in the pilot house. On unmanned barges, a cargo information card is mounted in a cylinder near a warning sign which indicates that the barge contains dangerous cargo. Each of these shipping documents has the shipping name of the product as well as an indication that hazardous materials are contained. However, it is important to remember that some product shipping names can be a broad generic category such as flammable liquid N.O.S., which means flammable liquid not otherwise specified. In this particular case, the specific chemical name will not be obtained from the shipping documents and other sources must be sought for assistance.

On some rail shipping documents, an additional number can be found which is known as the standard transportation commodity code. The STCC is a seven-digit number assigned by the railroads and used throughout the United States. If the rail waybill contains this number and the first two digits begin with 49, it indicates that the product is a hazardous material. There is a reference text available from the Bureau of Explosives, Association of American Rail Roads, 1920 L St., N.W., Washington, D.C. 20036, entitled "Emergency Handling of Hazardous Materials in Surface Transportation," which provides a cross-reference from the seven-digit number to a specific commodity name and a recommended action guide for emergency response personnel.

Also on the shipping document, and possibly on a placard for bulk transportation, is the four-digit United Nations number. This number assigned by the Department of Transportation to all regulated hazardous chemicals shipped in commerce provides the specific name of the product. The conversion from the four-digit code to the specific commodity is contained in the Department of Transportation book, "1980 Hazardous Materials Emergency Response Guidebook." Copies of this reference guide are available from the International Association of Fire Chiefs, 1329 18th St., N.W., Washington, D.C. 20036. Remember, the four-digit number will be shown on the placard only in certain instances when bulk cargo is being transported.

Employees can help

Plant personnel can be useful in providing information on products stored or used in processes in their facility. Usually up-to-date inventories are maintained. However, for early morning incidents when plant personnel are not usually available, it may be some time before this specific information can be obtained. However, the first-arriving chief officer must make every effort to determine the specific name of the commodity.

As indicated in one of the preceding paragraphs, preplan information specifically for maintaining a list of the hazardous commodities stored, utilized, or transported through a community can provide immediate information. It is difficult at off-duty hours to try and obtain this information from other sources. Therefore, a technique for preparing, maintaining, and updating this preplan information needs to be developed.

Finally, markings can be used in some instances to determine the specific commodity. Certain products must be identified in 4-inch-high letters on rail cars. For example, liquid chlorine rail cars, in addition to having a placard, must say that it is liquid chlorine on the outside of the rail car. The Department of Transportation requires these markings on 43 different chemicals. In addition, there can be markings with the name of the chemical on the outside of the shipping container. For example, a propane cylinder in addition to being marked a flammable gas can have a label on it that says propane. It is, therefore, necessary for the first-arriving chief officer to determine if any markings are visible. Remember that getting too close to read the labels may present safety hazards to the personnel under your command. A pair of binoculars is extremely useful in assisting in this step of the decision-making process.

Product hazard

Now that the product has been specifically identified, it is necessary for the first-arriving chief officer to determine the specific hazards which this chemical presents. Because there are over 35,000 different hazardous commodities produced by chemical companies, it is impossible for an individual to know how to handle each and every one. In fact, many of the major chemical companies have individuals who are expert in only three or four specific chemicals manufactured. They, too, find it impossible to know everything about all the chemicals produced by their own companies. It is, therefore, necessary for the first-arriving officer to be able to obtain information on the specific hazards from other sources.

Input for this decision-making step must come from:
1. Reference books.
2. Technical resources.
3. CHEMTREC.
4. Manufacturer.
5. Shipper.
6. Carrier.

There are many different reference books available for use at the scene of a hazardous incident. Two of these have already been mentioned. The key to selecting a good reference book involves the immediate information which can be provided to the first chief officer. However, care must be exercised to ensure that too many reference texts are not carried. Otherwise the chief officer will find himself pulling a large bookmobile behind the command vehicle and researching the chemical in several different books, finding all kinds of conflicting and differing data. Several texts which might be considered by the chief officer for reference include:

1. The small-sized edition of Volume I of the CHRIS Manuals from the Coast Guard that is entitled "A Condensed Guide to Chemical Hazards." It is available from the Superintendent of Documents, U.S. Government Printing Office, Washington, D.C. 20402.

2. The National Fire Protection Association's "Fire Protection Guide on Hazardous Materials," available from the National Fire Protection Association, Batterymarch Park, Quincy, Mass. 02269.

3. "The Fire Fighter's Handbook of Hazardous Materials" by Charles Baker, available from Maltese Enterprises, Inc., P.O. Box 34048, Indianapolis, Ind. 46234.

4. The "Farm Chemical Handbook" is very useful for fire departments which protect agricultural areas and could possibly be involved with the release of these types of chemicals. It is available from Meister Publishing Company, 37841 Euclid Ave., Willoughby, Ohio 44094.

5. For a guide to the emergency medical service problems and treatment of exposure to hazardous material incidents, the chief officer should reference the NIOSH "Pocket Guide to Chemical Hazards," which is available from the U.S. Department of Health and Human Services, Public Health Service, Center for Disease Control, NIOSH, Cincinnati, Ohio 45226.

Other resources

Other information on the hazard can come from technical resources which have been planned in advance. For example, there may be individuals in your community who have working knowledge of many different types of chemicals. These individuals should be approached to determine if they are willing to provide assistance to you as the chief officer should an incident arise in the community. Telephone lists must be continually updated to ensure that they are current.

CHEMTREC provides a 24-hour, seven-day-a-week, toll-free number to provide information on the chemical hazards of specific products. CHEMTREC can be reached from the continental United States by dialing 800-424-9300. From Alaska and Hawaii CHEMTREC will accept collect calls by dialing area code 202-483-7616.

To provide assistance, the CHEMTREC communicator will need certain basic information from the caller. In addition to the product name, information on the shipper, manufacturer, container type, the rail car number or truck number, the carrier name, a description of the problem, and a location of the incident will be necessary. Once given this basic information the communicator will provide the fire service caller information on the specific product. To assure accurate transmission of information both to CHEMTREC and back from CHEMTREC to the field incident commander, the fire department should establish a standard set of forms which include the data to be given to CHEMTREC and the format of the data which CHEMTREC will provide. Sample copies of these forms can be obtained by writing the author.

The manufacturer, shipper, and carrier are places that provide information on the specific hazards and suggestions on how to handle the incident. Some manufacturers, shippers, and carriers also maintain response teams which can be activated through CHEMTREC for direct assistance on the scene. In addition, several chemical groups have developed interagency response teams so that the nearest chemical company will respond to an incident even though it is not their product which is posing the risk. For example, the chlorine manufacturers have organized a group called CHLOREP. This team, activated through CHEMTREC, will send the closest available response group for a chlorine incident no matter who the original manufacturer or shipper of the product is. The individual response teams from these groups contain knowledgeable individuals who will come to the scene to provide assistance and advice to the incident commander. The only difficulty will be in the time it takes them to get to the scene.

Establishing objectives

Now that the incident commander has determined the specific product involved in the incident, the objectives for mitigating the incident need to be established. The areas of concern for the initial arriving chief officer involve:
1. Rescue.
2. Exposure protection.
3. Evacuation.
4. Water supply;
5. Containment.

As in all incidents which the fire service responds to, rescue is the primary objective. However, because a hazardous material incident is special, the incident commander needs to make several evaluations before rushing to effect a rescue. If the product presents severe health hazards and special protective equipment is required but is not available, should the risk of endangering additional personnel be taken? Should the incident commander ask himself, "If by attempting the rescue, will matters be made worse and will the individuals involved become an additional part of the problem?"

It is necessary for the first-arriving chief officer to face the fact that there will be some incidents where rescue efforts cannot be made until special protective equipment is obtained. This is a judgment decision which becomes extremely difficult. While some risk-taking is necessary, is it justified to risk the fire fighters in a given circumstance? The special protective clothing necessary may be obtained from outside resources or from a local specialized hazardous response team.

Hazardous material incidents also create specialized emergency medical service requirements. If the EMS service is part of the fire service and under the command of the initial chief officer, then what specialized equipment and protective clothing do these individuals have available?

During the treatment of personnel exposed to hazardous chemicals, will additional exposure occur to the EMS personnel? Will they, too, become part of the problem and not part of the solution? What specialized care is necessary for those victims who have been exposed to the chemical? Have the

EMS personnel obtained the necessary training to handle this specialized kind of problem?

Rescue needs

Rescue may also involve the need for specialized heavy-duty rescue equipment. As in the handling of the emergency medical service problem, these individuals also need specialized protective clothing for working in a hazardous environment. Training with the equipment is an absolute essential for handling a major accident involving the release of chemicals.

As in structural fire fighting, exposure protection is one of the major concerns of handling a hazardous material incident. To establish an objective for the protection of any exposures, the first-arriving chief officer must answer the following questions:
- What are they?
- Where are they?
- How can they be protected?
- What are the risks of protecting them?

In answering each of these questions, the incident commander can then establish specific objectives for protecting exposures. Remember that it may be necessary to provide for life safety and removal from exposures without the capability for preventing damage to them at some future point in the incident.

Evacuation decisions are very complex. Initially, the first-arriving chief officer needs to determine the immediate danger area. Civilian personnel must be removed quickly before the incident escalates. In addition, a further evacuation of the vicinity may be necessary based upon the circumstances of the incident. Suggested evacuation distances are contained in the Department of Transportation "1980 Emergency Response Guidebook."

Evacuation questions

The questions which will lead to the establishing of objectives for the evacuation requirement involve:
- Who will accomplish it?
- Where will the evacuees go?
- How will the evacuees get there?
- How will you handle the nonambulatory evacuees?
- Who will care for the evacuees?

Another requirement before the objectives can be established is to determine what water supply is available. While some chemicals are reactive with water, it is still the best means for providing for mitigation of the hazardous material incident. The questions the first-arriving chief officer must answer are:
- Where is the water located?
- Is there sufficient apparatus to provide water from the source to the scene?
- Is there sufficient water available from the source to handle the incident?
- Is a back-up water source available to ensure continued supply?

The final item which needs to be considered to establish the objectives involves containment. A hazardous material incident needs to be contained to prevent the escalating of the problem. Containment can be accomplished if the chief officer answers the following:
- How will containment be accomplished?
- What equipment is necessary to contain the spill or leak?
- How long will it take to get the equipment to the scene?

Alternatives

With the objectives established, the first chief officer must review the alternatives for solving these objectives. Input for determining the alternatives will come from several sources. These include:

1. The type of incident (spill, leak, or fire).
2. The state of the material (solid, liquid, or gas).
3. The hazard of the material (flammable, health, explosive, radioactive, reactive).
4. The terrain (distance to exposures, natural conduits, man-made conduits).
5. The time (slowly by a spill or leak; rapidly by explosion).
6. Life hazard (high, medium, low).

With this kind of input, the chief officer must determine the alternatives for controlling, extinguishing, or protecting life and property while withdrawing and evacuating. The techniques then used for the various alternatives will depend on all the information gathered in the decision-making process to this point.

Control can be accomplished in many ways. The spill, if it is a solid, can be covered with sheeting while a water-soluble flammable liquid can be diluted with water. In fact, some material may even be picked up, or if stored in containers, removed from the hazard area. Diking can be accomplished with sand, dirt, or special diking material available from various suppliers. Foam can be used to cover flammable liquid spills to reduce vapors and, in fact, special chemicals can be used to reduce the flash point or to convert a liquid into a semisolid, less-likely-to-run material.

If extinguishment is chosen as the alternative, the correct agents must be used. Vapor dispersion must be accomplished if a flammable liquid is involved. In addition, before extinguishment occurs, the chief officer must assure that there will no longer be continued leakage of vapor.

One of the hardest alternatives to select is to protect life and property while withdrawing an evacuation. This temporary holding action is foreign to many fire service personnel who have spent entire careers in an action-oriented position. However, as indicated previously, a hazardous material incident is unlike a structural fire, and it may be the best course of action to select this alternative.

Once all of the evaluation has been done, the objectives set and the alternatives considered, the chief officer must then make the final selection. Which alternative is best for the given incident? Certainly in the first minutes of arrival on the scene of a hazardous material incident, the information continues to flow to the commander. On the basis of the information immediately available, the best alternative is selected. However, remember that this is a continuing evaluation process, and as more information is received, as the alternative selected proves not to provide the best solution, reevaluation must take place.

Reconsider alternatives

Based on the reevaluation, another best alternative can be selected. Remember, making a change is not an error. It is an important part of the process which must be considered. Chief officers should not put on blinders, make a selection, and go down the road toward a solution without ever reconsidering that decision. A hazardous material incident is a very dynamic situation which continually changes. No one decision should be considered "the answer."

When the incident is concluded, remember that cleanup of personnel, apparatus, and equipment must be accomplished. It would be a shame to have successfully handled an incident with all the specialized protective equipment necessary without injury and then return to the fire station and have the fire fighters exposed to the chemical residue while handling their clothing or apparatus. This is an important step in the entire decision-making process which cannot be eliminated.

The first-arriving chief officer has a great deal to think about. A lot of decisions must be made quickly. Information needs to be provided to the command post quickly and correctly. Only with correct and good information will the incident commander be able to make decisions at a hazardous material incident. These decisions need to be based upon a systematic analysis and must be continually practiced and worked with if they are to be successful. A haphazard procedure can lead only to injury, confusion, problems and an escalation which will harm the community. Be prepared to handle the hazardous material incident in your jurisdiction. ☐ ☐

ADD AN EMS SPECIALIST TO YOUR HAZ-MAT TEAM

Hazardous material response teams are good for any community, but team personnel take a tremendous risk. EMS specialists provide an extra degree of assistance for everyone.

Montgomery County, Md., has the potential for serious incidents involving many dangerous substances. Within the county, there are two interstate highways, a main line of the Chessie System Railroad, five major pipelines, two small plane airports, air traffic to and from nearby major airports, thousands of businesses selling or storing chemicals, and many medical and military facilities. These facilities contain materials including etiologic agents, radioactive materials, explosives, flammable liquids, oxidizers and poisons. In addition, there are mutual-aid agreements with jurisdictions throughout the Washington, D.C., metropolitan area.

The Montgomery County Fire and Rescue Services Hazardous Incident Response Team (HIRT), which became operational on Nov. 1, 1981, has recognized the preceding problem and included six EMS and medical specialists as part of the team. This special EMS group is composed of four paramedics, a registered nurse who is also the EMS officer, and a physician. Other team members include four shift officers, 13 fire fighters, four fire fighter trainees, a supervisor and two coordinators.

HIRT, whose members serve on a volunteer basis, provides support services to the 18 independent fire and rescue departments in the county. The team operates under a rotating shift schedule, with

BY CHIEF WARREN E. ISMAN
and
SCOTT GUTSCHICK
Department of Fire and Rescue Services
Montgomery County, Md.

The Montgomery County, Md., Hazardous Incident Response Team vehicle is an altered surplus minipumper that carries a wide assortment of equipment.

two shifts on standby and two additional shifts on alert at all times.

All applicants for membership on the team must be 18 years old, complete a total NFPA 1001 physical examination, have three years of fire service experience, be a certified NFPA 1001 Fire Fighter III, be a certified State of Maryland EMT-A, and have completed the Montgomery County Fire Academy hazardous materials course.

Having met these requirements, new personnel are assigned to shifts as fire fighter trainees. They assist at incidents in every manner except the handling of the actual hazard. Once the individual has obtained the required knowledge and skill level, the trainee is promoted to team member.

An EMS specialist is assigned to each HIRT shift. This EMS specialist, in addition to the requirements outlined above, must be a Maryland Certified Paramedic (cardiac rescue technicians) or paramedic instructor.

The team is also fortunate to have an emergency department physician who serves as the HIRT medical director. His assignment was approved by the EMS committee of the Montgomery County Medical Society, which provides medical direction to the prehospital advanced life-support program in the county. Another of the team's medical specialists serves as a consultant to the U.S. Surgeon General on emergency medical services and assists the team as a medical advisor and as a paramedic.

In terms of emergency care, the EMS specialist must always be ready to administer treatment with regard to the specific hazards involved. Hazardous materials can produce serious and sometimes unusual injuries such as frostbite, cellular dysfunction, poisoning, paralysis, chemical burns, infections and asphyxiation. Unless countermeasures are taken immediately, permanent injuries or death can occur. For this reason, the HIRT vehicle carries a complete drug kit with many special drugs that are capable of treating unusual symptoms caused by certain hazardous substances.

The drugs which are carried, based upon the recommendations of the HIRT medical director, include: 50 percent magnesium sulfate, dexamethasone sodium sulfate, amyl nitrite capsules, 0.5 percent tetracaine hydrochloride, 1 percent methylene blue, and 0.9 percent sodium chloride. The EMS compartment also contains a first-aid kit, resuscitator, two cyanide treatment kits, two bottles of eye wash solution, four bottles of normal saline solution and an EMS reference library.

The HIRT uses a surplus mini-pumper that has undergone significant alteration by team members. Modifications have included removal of the tank and pump, addition of a floor, compartment shelves and weather cover, and a complete paint job. The vehicle is stationed at the county's public service training academy where it is centrally located and close to Interstate 270 and the Chessie System Railroad lines. The responsibility for driving the unit rests with the fire and rescue training staff and designated HIRT members.

The HIRT vehicle also carries a complete drug kit with many special drugs capable of treating unusual symptoms caused by hazardous substances.

In order for the team to be prepared for incidents involving a wide variety of substances and containers, the HIRT vehicle is equipped, in addition to the EMS materials, with a large assortment of specialized protective gear, leak control materials, suppression equipment, monitors and tools. These items have been obtained by means of county government funding, donations from the private sector and the ingenuity of the team members.

The team has also acquired the use of a spill response trailer from the state's water resources administration. It can be quickly hitched to the HIRT vehicle when needed at the scene of a large spill. The trailer contains various sorbent materials, recovery drums, fencing, nets and thick plastic bags.

Equipment that the team has found to be most useful during its responses include 30 and 60-minute SCBA, particle filter masks, sorbent C, spill control pillows, acid neutralizer, "No-Flash" neutralizer, recovery drums, explosimeters, air sampler monitors, Geiger counters, shovels, brooms, hand tools and reference materials. A complete inventory of the equipment carried is available.

EMS Functions at Hazardous Materials Incidents

Pre-Incident

- Development of safety operation procedures.
- Development of basic and advanced life support protocols specific for hazardous materials incidents.
- Development and maintenance of resource list for medical consultation.
- Development and maintenance of first-aid and medic kits for the HIRT.
- Identification of possible hazardous materials incidents with the potential of major medical problems.
- Development and maintenance of base line (normal) health record system for HIRT members.
- Participate in regular drills and meetings, and assist in the maintenance of equipment.
- Training of HIRT and other fire and rescue personnel on safety procedures, protective gear, decontamination, health hazards and EMS matters related to hazardous materials incidents.

On the Scene

- Serve as health and safety officer for team members.
- Serve as advisor on emergency medical matters to the HIRT officer and incident commander.
- Maintain exposure records. The cumulative effect, even with the use of protective equipment needs to be watched. In addition, concern for future illness, which may be many years developing, needs to be handled. For this reason, an incident exposure record is maintained at the scene showing all of the chemicals to which the individual has been exposed. This information is then transferred to the personnel exposure record and kept in the individual's permanent file.
- Provide initial emergency care for HIRT members.
- Assist field EMS personnel with providing care for specific health problems.

Post-Incident

- Coordinate decontamination activities.
- Assist with conducting critique of incident.
- Coordinate follow-up care of personnel exposed to hazardous materials.

During its first nine months of service, HIRT has assisted on 31 incidents involving such materials as pesticides, propane, anhydrous ammonia, hydrochloric acid, sodium hypochlorite, diesel fuel, fuel oil, gasoline, sodium arsenic, cyanide and acetone. The team has also been called upon to assist with determining types and concentrations of suspected toxic vapors.

HIRT members report directly to the scene of an incident from their homes, work or fire stations under their own means of transportation. Alerting of team personnel is done by the county's emergency operations center with tone-activated pagers. Special arrangements have been made between the HIRT and the employers of the team members so that HIRT personnel can be released to respond to incidents when their shift is on duty. Arrangements have also been made with the county police department to allow team members to pass quickly through road blocks. HIRT personnel can be identified by means of their special windshield identification, turnout gear identification tags and wallet-sized picture identification card. These individuals also wear fluorescent vests over their protective gear whenever practicable.

Although team members are experienced and knowledgeable about the handling of hazardous materials, additional training takes place on a regular basis. Team drills are held at least monthly and are usually conducted by the HIRT personnel themselves. This is accomplished by each shift being responsible for various topics which must be researched before providing the instruction. Experts from both the private and public sectors also participate on occasion.

The four shifts also drill separately one or more times per month at the discretion of the shift officer. When funds are available, individuals attend special training courses and conferences.

The officers, coordinators and supervisors hold a monthly meeting to discuss previous incidents and to plan future activities. In addition, the entire team meets once each month to keep everyone informed on new developments and to assign new tasks.

HIRT participates in many hazardous materials programs, too, in order to keep ideas and techniques flowing to and from the team. Team members give many presentations on handling hazardous material incidents to local, state and national conferences. The team also has plans to cosponsor a seminar on EMS response to hazardous materials emergencies and to participate in a training program for fire department officers from the Washington, D.C., area.

Although the HIRT has the capabilities to handle most of the incidents it is called upon to handle, the team is not reluctant to call in outside assistance from competent individuals and organizations. The team maintains a resource manual that lists the local, state and federal agencies that can be contacted for assistance. It also lists the resources available from the private sector, as well as the names and numbers of technical assistance organizations.

The team is always looking to improve its efficiency so that it can better protect the 600,000 citizens of the county. Projects that are currently under way include the acquisition of pre-incident plans for target areas, a skills checklist and decontamination procedures. HIRT has 15 plans so far and is working on many additional ones.

A skills checklist is being developed so that HIRT officers will know the specific capabilities of each team member. This will assist the officer in assigning tasks during incidents.

Individuals desiring a list of the HIRT inventory, the resource manual outline, forms or any further information about the team should contact Chief Warren E. Isman, Director, Department of Fire and Rescue Services, 101 Monroe St., 12th floor, Rockville, Md. 20850. ◻ ◻

DELAYED ALARM AT CHEMICAL FIRE

A warehouse employee detected fumes from a burning chemical in an area where 700 to 800 250-pound drums of sodium hydrosulfite, a textile bleach, had been stacked an hour earlier. But instead of calling the fire department, he went across the street to a restaurant and called his company's main office. The owner of the company then drove six blocks to the warehouse with breathing apparatus. Heavier fumes were encountered, and dry chemical extinguishers obscured vision. Finally, a decision was made to call the fire department. This was done by walking back to the restaurant and calling the company office again. A secretary there dialed 911.

BY BRADLEY ANDERSON
Fire Inspector
Charlotte, N.C., Fire Department

When the Charlotte, N.C., Fire Department got the 911 call to a chemical fire last Sept. 13 at 1611 hours—after about a 20-minute delay—it dispatched a box alarm assignment. Within two minutes, Engines 11, 4 and 7, Ladder 4 and Battalion Chief Luther Fincher arrived to find gray smoke showing at the rear of the one-story brick structure. Fincher established command at the loading dock on the northeast side of the building. Because the secretary reporting the fire had identified the product involved as sodium hydrosulfate, the fire alarm center had relayed hazardous materials data to the responding companies.

The companies were informed that sodium hydrosulfite is water reactive—that dry chemical, carbon dioxide or sand are the best extinguishing agents and that water should not be used unless flooding amounts could be applied. Personnel were informed that SCBA should be worn for protection from sulfur dioxide and hydrogen sulfide released in the combustion process. These products combine with moisture of the skin, mouth, eyes and nasal passages to form sulfuric acid.

In sizing up the situation, Fincher confirmed that a drum of sodium hydrosulfite was smoldering. (Fire investigators later learned that it had been damaged, exposing the chemical to moisture in the air.) Now employees wearing SCBA were attempting to move 1000-pound pallets of sodium hydrosulfite. These pallets were stacked to the ceiling. A forklift truck was being used to gain access to the smoldering drum.

Fincher ordered Engine 11, Engine 7 and part of Engine 4's crew to report into the fire scene as Engine 4 stood by the hydrant in front of the building. Fire fighters entered the warehouse with dry chemical fire extinguishers in an attempt to prevent fire development, remove the hot drum from the building and isolate it from the other drums.

At 1618 Fincher was advised by crews inside the building that they could not locate the involved drum due to poor visibility and tightly stacked pallets with no aisles.

Exhaust fans were set up to remove fumes and dry chemical powder. Truck 7, a heavy rescue vehicle with additional large fans, generator, lights and air cascade, was special-called. Efforts continued to remove the drum, but at 1624 Engine 4's crew reported flames visible. This was quickly followed with a report that the fire had been extinguished. Fire alarm, overhearing the report, cautioned Engine 4 that the hazardous materials information indicated sodium hydrosulfite can reignite in the air. Engine 4 acknowledged the caution and reported that the product had, in fact, just reignited and that flames had been extinguished a second time with dry chemical.

Dry chemical extinguishers were used repeatedly to extinguish the fire which persistently reignited after each application. Fire fighters continued to work at removing other drums to get to the involved drum.

Second alarm, new tactics

Fincher ordered a second alarm at 1633 hours, because dry chemical and carbon dioxide extinguishers on the scene were depleted. Engines 1, 5 and 2, Platform 1 (an 85-foot aerial apparatus with articulating boom and basket), a hose tender, Squad 1 (a flying personnel squad) and Battalion Chief Roger Weaver responded with additional extinguishers. A special alarm followed for Blaze 7, a quick-response airport crash truck equipped with 450 pounds of dry chemical, but tactics had to be changed before it arrived.

By 1657 hours conditions had worsened to involvement of 15 or more drums. Flames from floor to ceiling became visible to crews working inside the structure. All personnel were ordered out of the building as drums began to rupture.

The offensive mode had to be changed to a defensive one. A high-volume attack was prepared.

The door through the parapeted fire wall was closed to prevent extension to front areas of the warehouse. Engine 4, which now had a 2½-inch hand line laid to the loading dock for backup protection of personnel, set up a master stream appliance in this area supplied by three 2½-inch lines. Rush-hour traffic was rerouted as Engine 11 made a forward lay of 4-inch supply line from the corner of Keswick Ave. and Dunlow St. to the rear of a trailer parking lot adjacent to the east corner of the warehouse. Engine 5 made a steamer connection and relay pumped.

Platform 1 was set up on the west corner of the warehouse, supplied through a 4-inch line from Engine 2 at the corner of Sylvania Ave. and Tryon St. From this position the platform could protect the two-story, metal-clad offices of a neighboring feed processing plant separated from the fire by a 4-foot-wide alley. This protection was supplemented by an exterior sprinkler system on the side of the office structure.

A special alarm was ordered for Ladder 2, while Ladder 4 set up along the railroad tracks at the rear of the structure. It was supplied by a 4-inch line hand laid from Engine 11 through a fence and heavy brush.

As this operation was being set up, fire fighters, with the assistance of police, were evacuating the feed processing plant and railroad yard on either side of the warehouse. Acrid fumes were making these areas unsafe for persons without SCBA protection. Weaver, who had command of the rear of the building, reported brown smoke visible, indicating structural involvement.

Simultaneous deluge planned

Ladder 4 was instructed to begin dispersing vapors with the water tower stream but to avoid application of water to the fire building until other companies were ready to simultaneously deluge the sodium hydrosulfite.

Because drums were stacked tightly in a confined area that prohibited an interior attack, it was decided by Fincher, conferring with Assistant Chief R. L. Blackwelder who had arrived on the scene, that the only way to ensure a high volume attack in the right place was to let the roof burn away. This would allow master streams to simultaneously hit the seat of the fire and flood all the drums of chemical.

No one was sure how the sodium hydrosulfite would react or what amount of water would be adequate to flood such a large quantity of the chemical. Personnel were instructed to keep as much distance between themselves and the building as possible.

Paraquat discovered

Matters were complicated as fire fighters withdrew — retrieving a sample one-gallon plastic jug of liquid which Fire Marshal Art Goldner identified as paraquat. Chemical data on the herbicide could not be found in any of the hazardous material guides on the scene or at fire alarm. Not knowing its properties, officers used extra caution to keep fire fighters out of the smoke, even those wearing SCBA, until the manufacturer was contacted through CHEMTREC and verified that paraquat could not be absorbed through the skin. Fire alarm was advised that heating of paraquat releases carbon monoxide, toxic oxides of nitrogen and sulfur, and hydrogen chloride, which combines with moisture of the body to form hydrochloric acid.

In all, 3200 gallons of paraquat were present in the fire area. The company owner also told fire fighters about additional 800 cases of tobacco insecticide in the building.

All personnel exposed to smoke were ordered to operate with full protective clothing and SCBA and the command post was moved north on Tryon St. Later this position became untenable without SCBA, causing the command post to be moved farther along Tryon.

Engine 7 laid a 4-inch line from the hydrant at 24th St. and Tryon to Ladder 2 at the north corner of the building. By 1735 a large hole had burned in the roof. Companies were ready to flow water and personnel were instructed to position master stream nozzles and then withdraw in the event of an explosion. Fincher ordered the attack to begin.

Deluge tactic works

The booms of Ladder 4 and Platform 1 were turned from exposures to the fire it-

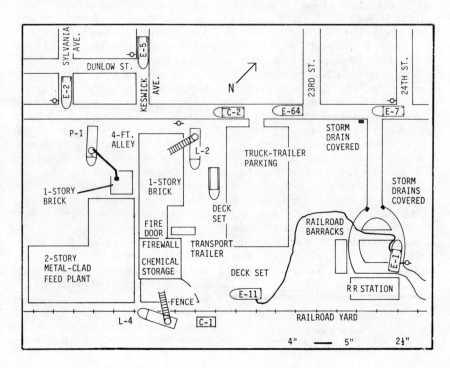

Chemical fire . . .

self. Engine 4's throttle was advanced to supply the master stream appliance on the east side of the building. Engine 7 and Ladder 2 also revved up their pumps but a section of hose burst in the 1500-foot lay, putting Ladder 2 out of commission for several critical minutes.

Nevertheless, the deluge tactic worked. The main body of fire was controlled. Portions of the roof had collapsed, however, blocking fire streams from several hot spots, and drums continued to rupture.

To supplement Ladder 4's water supply, the hose tender laid a 5-inch line from a hydrant at the railroad station to Engine 11, with Engine 1 relay pumping. This also enabled Engine 11 to pump into a master stream appliance at the rear corner of the warehouse.

Engine 7 continued to have bad luck. The burst section of hose was replaced and water flow resumed at 1744. But in two minutes a coupling blew apart, striking a fire fighter. Medical aid was administered as an additional engine was called to relay pump the line. Engine 64 was in position in 20 minutes.

When the fire first ventilated the roof, smoke rose above the ground on a thermal column, carried on a mild easterly breeze. The only threat of smoke inhalation was to personnel in the immediate area of the building.

This condition quickly changed. Shortly after the exterior attack began, wind direction changed 180 degrees and cool evening air settled to the ground, holding smoke down as it was pushed into the residential neighborhood west of the incident.

Immediate evacuation of the neighborhood was necessary. Four city buses were pressed into emergency service, transporting 600 to 800 residents to shelters.

At 1903 Blackwelder assumed overall command of the incident, with Fincher as fireground commander, assisted by Weaver in command of the east sector and Battalion Chief Bill Summers (who had reported from off duty) in command of the south sector.

Wind direction fluctuated in the next two hours. With smoke continuing to push west and north at ground level, the evacuation area was expanded several times. The city department of transportation altered the traffic signal computer program to expedite evacuation from the area.

There was a heavy demand for breathing air throughout the night. Not only were fire fighters heavily dependent upon air bottles (some 30-minute tanks were refilled 15 to 20 times) but police officers responsible for evacuation were given quick lessons on the scene in the use of SCBA by the fire department training division.

Truck 7's air cascade system was supplemented by cascade systems and extra SCBA brought to the scene by the Charlotte Life Saving Volunteer Rescue Squad and nine volunteer fire departments from Derita, Hickory Grove, Mallard Creek, Mint Hill, Newell, Pineville, Pinoca, Statesville Road and Steele Creek. Cascade cylinders were shuttled between the fire scene and the fire department supply division for refilling. In all, more than 50 300-cubic-foot air cylinders were used.

After flowing an estimated 4000 gpm for an hour and a half to cool reacting sodium hydrosulfite, master streams were shut down. Engine 11 stretched 1½-inch hand lines to the rear overhead door. The frame of the door was removed with the help of a winch on Brush 22, and fire fighters began the tedious overhaul process of removing what remained of the sodium hydrosulfite drums one at a time, flooding each with water. A front-end loader was brought to the scene to assist in the removal and to build a dike around the loading dock for containment of the chemicals.

Even as this operation was beginning, a drum ruptured. Fire fighters proceeded with caution, still wearing full protective clothing and SCBA. Overhaul continued well into the next day.

Environmental officials, automatically notified by fire alarm after arrival of the first equipment, anticipated a problem with contaminated water runoff from the fire scene. Salvage covers were placed over storm drains downhill from the fire scene in front of the railroad station to prevent runoff into the drains. These tarpaulins were held in place by sand hauled to the scene by the street maintenance department.

The method worked. Forty thousand gallons of paraquat-contaminated water were contained and later hauled away for disposal.

Unfortunately the reservoir created by the blocked drains was insufficient to control all the runoff. Officials discovered

The master stream from Platform 1 helps flood the chemical. The deluge tactic worked.

Photo by Elmer Horton, The Charlotte News

that paraquat was entering Little Sugar Creek, which flows through the watershed several blocks northeast of the fire scene. To control this runoff, two earthen dams were constructed using sand trucks and a backhoe. The next day a third dam was constructed.

Following much concern over contamination of drinking water supplies downstream in South Carolina, environmental officials took a calculated risk, systematically opening the dams four days later. The calculations proved accurate and paraquat was diluted to safe levels as the creek mixed with the Catawba River before reaching water treatment plants.

Five fire fighters suffered injuries including smoke inhalation, bruised ribs, a facial laceration and contusion of an ankle. Over the next several days, 61 fire fighters suffering sore throats, skin rashes and flu-like symptoms were checked by the city employee medical services. Blood and urine samples were collected from 109 fire fighters for laboratory analysis to determine the extent of medical problems. No one is sure what the long-term medical effects of this incident will be.

Most of the first and second-alarm C-shift personnel were relieved at 1800 hours when A-shift came on duty. As prolonged operations continued, these personnel, wearing SCBA and protective clothing, were under considerable strain. Beginning at 2300 hours, engine companies from outlying stations were called to the fire scene, and their crews switched with engine companies committed to the fire. Crews from Engines 6, 18, 8, 22, 15 and 21 were matched with apparatus similar to theirs to minimize unfamiliarity. Relieved crews were returned to the outlying stations with the uncommitted apparatus.

Because each of the three pieces of aerial apparatus in use was unique, only the crews assigned to those apparatus were considered knowledgeable enough to safely operate them. These crews were transported to their respective fire stations for showers and a change of clothing, then returned to the scene for the remainder of the night.

At 0448 the next morning, the incident was declared under control. At 0530 hours evacuated residents were permitted to return to their homes. They were instructed by health officials to discard any food that had been exposed to smoke.

The chemical company hired a chemical salvage contractor to clean up the warehouse site under the supervision of the environmental health department and the regional EPA office.

The last engine company cleared the scene on Wednesday at 1630, although sodium hydrosulfite continued to smolder for another 24 hours.

This incident identified a weak point in our fire inspection program. The fire prevention bureau had ordered correction of fire and building code violations at the warehouse before the incident occurred. Violations included storage of chemicals without a fire department permit, failure to adequately separate the chemicals, pallets stacked too high, no aisles provided between pallets, failure to post 704M warning placards on the exterior of the warehouse (see ``Entire Buildings Are Labeled With NFPA 704M Marking System,'' Fire Engineering, September 1982), and a storage of these chemicals in an unsprinklered building.

Because the owner had been ordered to remove the chemicals, their presence was not updated on the building record form for the warehouse that is kept on file at fire alarm. Placards that normally identify hazards present at chemical incidents were not displayed either. Had the chemicals been stored in compliance, their presence and basic properties would have been recorded on the building record and posted in the form of 704M placards.

Proper storage of the sodium hydrosulfite with wide aisles and restricted height of pallets would have made locating and isolating the damaged drum much easier. A sprinkler system could have provided the flooding quantities of water in the proper place to cool exposed drums.

The incident is causing code enforcement policies within the department to be tightened up to allow less time for correction of violations and prompt legal action for noncompliance. At this writing the city is pressing legal action against the chemical company. ◻ ◻

Unstable Chemical Controlled Safely

BY WILLIAM RATHBUN
Battalion Chief
Ft. Lauderdale Fire Department

The value of learning the nature of specific hazardous chemicals stored at a manufacturing site—before an emergency incident occurs—was once again demonstrated. Called to a large plant that produces plastic lenses for the eyeglass industry, the Fort Lauderdale, Fla., Fire Department found smoking cartons of diisopropyl peroxydicarbonate being thrown into a small lake on the plant property last Feb. 23.

The material is stored in plastic containers in 10-pound blocks and kept at 70 degrees below zero in electric freezers. But a preheater on the lid of freezer 1 malfunctioned and burned the rubber insulation away from the lid. When outside air then entered, the temperature started up and the contents started decomposing.

The smoking chemical was not removed from the plastic containers when they were taken out. Five of the packages sank in the lake. These caused no further problems, but two floated on the surface and continued to give off invisible fumes which affected eyes and throats.

Thought under control

Inside, other decomposing blocks had been removed from the defective freezer and placed in another unit maintaining the required temperature of −70 degrees. Plant personnel indicated that the situation was under control, and the other fire department personnel left.

I was on the scene for about 15 additional minutes when a plant worker came running from the blockhouse and stated that the temperature in the second chest had started a rapid rise, gaining 20 degrees in 15 minutes.

I recalled the companies to the scene and ordered Captain Janson to start an evacuation of a two-block area north of the blockhouse because this building is designed to explode outward in that direction. The police department assisted in the evacuation.

Engine 2 was ordered to stretch two 2½-inch supply lines to a deluge gun set up at the front of the blockhouse. When the lines were in place, a long wait began.

At this time, Captain Robson and Lieutenant Remer were on the phone to CHEMTREC trying to get additional information on the proper course of action to take. CHEMTREC did not have any new information, but they did put us in contact with Gary Gardner in Barberton, who is the plant chemist where this material is being manufactured. Gardner told us not to try to move any of the smoking material because this was very dangerous. (This, of course, had already been done earlier by the plant personnel.) Gardner told us that if the remaining material had not reached the critical stage (which is somewhere around zero or slightly above), decomposition could be reversed with the addition of dry ice.

As soon as a source of dry ice was found, at a local dairy, Engine 9S made an emergency run to get a good supply. Meanwhile, plant workers wanted to dump the entire contents of the second chest—30 cartons—into the lake. Not knowing what the consequences would be (the first few without any reaction could have been just luck) and being concerned about environmental damage, we took time to discuss this further.

It was agreed that we would move most of the decomposing containers to a third freezer. When all men had protective gear ready, including SCBA, six of the blocks in the worst condition were removed from their cartons and sunk in the lake. The rest were moved to the new chest—until it also started to heat.

Dry ice helps

We feared we may have had an unstoppable reaction underway, so everyone was pulled back to the command post to wait on the dry ice. When it arrived, the temperature of the third chest was up to 20 degrees below zero—still in the safety zone but well above normal. One hundred and fifty pounds of dry ice were added to the top of the chest. All personnel pulled back again to wait.

The temperature continued to rise for 15 minutes, stabilized for 20 minutes and then started to drop. When the thermometer showed a drop to 50 degrees below zero, the material was turned back to plant personnel. The incident was under control 5½ hours after it began.

For next time

The following information should be helpful in the event of another incident. Getting past the chemical name is half the battle. There are three chemical names, all taken from the base chemical peroxidicarbonate:

- Diisopropyl peroxydicarbonate (IPP)
- Di-n-propyl peroxydicarbonate (NPP)
- Di-sec-butyl peroxydicarbonate (SBP)

All three chemicals have to be stored and shipped under refrigeration, at temperatures well below 0°F. IPP, shipped as a solid, melts at 46 to 50 degrees, and rapid decomposition occurs at 57 to 64 degrees. NPP and SBP, shipped as liquids, likewise decompose in this same temperature range.

If allowed to reach rapid decomposition temperatures, peroxydicarbonate gives off heat and flammable vapors. These vapors can easily catch fire from an ignition source or even ignite spontaneously from the heat generated by decomposition. If this heat is not removed, the decomposition becomes self-accelerating. Decomposition vapors are white or colorless until they catch fire. Keep ignition sources away.

Keep the material clean. Peroxidicarbonates are powerful oxidizers. Contamination by other chemicals, metals, dust or dirt can cause the material to decompose and give off flammable vapors.

If refrigeration fails, monitor the storage area holding pure peroxydicarbonate material and solutions visually, as well as with a high-temperature contact alarm set to give a warning signal well below the safe upper limit. Use dry ice to keep stored material cold. Place the dry ice above the peroxydicarbonate, not below it. If dry ice is not available, promptly dispose of the material.

Disposal

When using IPP, NPP or SBP, keep a container of sand or inert material such as vermiculite in the work area. Cover small spills with sand and dispose of the material immediately by taking it in a clean, open container to a well-ventilated area free of any combustible material. Scatter the contaminated sand on the ground and allow the peroxydicarbonate to decompose gradually.

A large quantity of waste IPP, NPP or SBP can be disposed of by scattering it on the ground in an open, well-ventilated area free of ignition sources and combustible materials.

However, based on the experience at our incident with the two floating containers and the amount of eye irritating fumes given off, caution should be used in spreading a large amount anywhere near a residential area. The fumes from 30 or more containers would cause a serious problem. It would be prudent to pack the material in dry ice and remove to a remote area for dispersal.

Don't fight the fire

In case of fire, leave the area as fast as you can. Don't try to fight a peroxydicarbonate fire. Let it burn itself out. Limit the fire by spraying water from a safe distance to cool surrounding buildings and equipment. If the peroxydicarbonate storage units are not involved in fire, spray them with water from a safe distance. Don't spray peroxydicarbonate trays or bottles directly. Don't start cleanup and salvage operations until the area has cooled down completely.

Remember: A fire may lead to an explosion. Keep a safe distance from a fire. A distance of 2500 feet is recommended as a safe initial distance.

Any of the situations described that permits peroxydicarbonate material to decompose could also cause an explosion. This is the reason: If the vapors formed by decomposition cannot readily escape, they will build up pressure and burst the walls of a container. Storage containers of all types (bottles, catalyst charge vessels, storage tanks or buildings) should permit ready release of vapors resulting from decomposition.

Peroxydicarbonates become more sensitive to impact as temperatures increase. Tests made by the U.S. Bureau of Mines showed that certain severe conditions of impact plus confinement can cause explosive decomposition. Do not skid or drop peroxydicarbonates. IPP is not considered to be sensitive to friction, but you should avoid unusual conditions of friction such as grinding.

Consult a physician

These chemicals, both pure and in solutions, are extremely irritating to the skin and eyes. Adequate ventilation is necessary.

In case of contact with the skin, flush the contaminated area with water. Clean with soap and water and then with rubbing alcohol, if available. Take a shower. Remove contaminated clothing, let it hang out to air and launder before using again. Consult a physician. In case of contact with the eyes, flush them with water for at least 15 minutes, and consult a physician immediately.

For help in emergency situations with these chemicals, call PPG Industries plant at Barberton, Ohio, (216) 753-4561. The chief chemist is Gary Gardner.

Gardner stated that there should be no chemical reaction if the chunks are immersed in a body of water, but they should be placed at different locations and not all dumped on top of each other. The material should break down into isopropyl alcohol and not harm the water. But dumping in water should be a last resort if all other efforts fail. Nor should the material be flushed into the sewer system.

Editor's Note: According to the manufacturer, PPG Industries in Ohio, approximately 100 plants store large quantities of diisopropyl peroxydicarbonate for use in making plastic lenses for eyeglasses. The plants are mostly on the East and West Coasts.

Upwind hose streams and elevating platform help disperse the vapor cloud without hitting the tank, which would have added relative heat.

Plan Works, Leak Stopped In Liquid Nitrogen Tank

BY CHIEF PAUL E. ALBINGER, JR.
Saukville, Wis., Fire Department

After receiving a report of a large amount of smoke at the Freeman Chemical Corp., the Saukville, Wis., Fire Department responded and found vapors leaking from a pressurized tank of liquid nitrogen.

The tank measured about 30 feet in height and 12 feet in diameter. It was located in a diked tank farm containing about 30 tanks of various types of polar solvents and petroleum-based chemicals.

This call, around 3 a.m. Sunday morning last Aug. 31, came on the heels of a July 25 call when an acid tank of thallic anhydride caught fire after being struck by lightning. That incident forced the evacuation of over 150 area residents.

Most important

We feel the most important part of handling this incident occurred years ago when a pre-fire plan was developed for Freeman Chemical. The company is a large manufacturer of synthetic resins, foams and polyesters and it is Saukville's largest industry. The pre-fire plan consists of a three-alarm, 10-fire department response plus a special call of the Ozaukee County Rescue Squad specialty unit which handles any necessary evacuation. The first alarm calls for an automatic response from the Port Washington Fire Department.

For this incident, the Saukville Fire Department responded with Engine 365, a 1000-gpm attack pumper; Engine 366, a 1000-gpm pumper; Engine 364, a 1500-gpm pumper (which was placed in service only one week prior and was on its first call); Unit 356, an equipment truck and command post; and Squad 352, an ambulance. The Port Washington Fire Department responded with Engine 461, a 1500-gpm pumper; Engine 462, a 1250-gpm pumper; and Unit 460, an 85-foot elevating platform with a 1000-gpm pump.

Stop and think

The pre-fire plan has been practiced and used many times. Frequency and training gives everyone knowledge and experience on standard operating procedures. Radio communications are kept to a minimum and confusion is avoided. The most important thing we did was to stop and think. A spontaneous action could endanger the lives of our fire fighters. The incident posed enough problems without our becoming a part of it.

Our personnel who have completed hazardous materials training took our DOT Hazardous Materials Guide and looked up liquid nitrogen. It was listed as a nonflammable gas. It is poisonous as a concentrated vapor cloud and seals off oxygen.

I live two blocks from Freeman and upon my response the vapor cloud had enveloped two streets. Liquid nitrogen is also very cold in nature. This caused us major problems because the vapors being expelled from the tank formed a large layer of ice and frost over the valves that needed to be closed. Also, if water were to be directly applied to the tank we would actually have added relative heat to the tank and liquid nitrogen, causing even more leakage. It is of utmost importance to secure proper information about an incident before acting.

Upwind approach

All units were ordered to approach the incident upwind. We were concerned about the dispersal of the vapor cloud.

Engine 365, our attack pumper laid a 4-inch line from a preselected spot. For 365's water supply, Engine 364 then laid its own 4-inch line from 365's split lay to a plant hydrant supplied partially by the municipal and Freeman's own system. Engine 365's crew of eight men stretched a preconnected 2½-inch hand line and two 2½-inch lines to a deluge set with a fog nozzle to start dispersing the vapor cloud.

Engine 366's assignment is to enter the rear entrance of the plant and then lay parallel 2½ and 3-inch lines to a hydrant on Linden St. adjacent to the plant and supply another deluge set. However, a call for a house fire was received upon our arrival at Freeman, and Engine 366 could not be committed at this time. Unit 356 set up as a command post, and Rescue 352 stood by as an EMS unit. It left to answer another call at 3:35 a.m.

Water relay

Port Washington's Engine 461 arrived and reported to a hydrant on Tower Drive with a 12-inch main. Three hundred feet of 4-inch line was laid from a hydrant that feeds a main running to Freeman's plant system. Then 461 pumped from the municipal water system into this hydrant and relayed to Engine 364 to supplement its supply.

59

Photos by Jim Carrier, Saukville F.D. photographer

After the valve is defrosted, fire fighters in full protective gear shut off the leak.

Engine 364 then relayed to 365. This unique system is similar to pumping into a sprinkler siamese and can yield a combined flow of over 2000 gpm.

Unit 460, the 85-foot elevating platform, set up upwind near 365 and operated a 750-gpm stream using a fog nozzle. Engine 462 laid 600 feet of 4-inch hose to a hydrant on Rail Rd. and Church St. and supplied Engine 460. This operation was successful in totally dispersing the vapor cloud and eliminated any further need of evacuation. Our actions could now be turned to stopping the leak.

In the meantime, we had notified Paul Schaefer, Freeman Chemical plant manager, and he had arrived at the plant. The tank was owned by Airco Industrial Gases of Milwaukee. Liquid nitrogen is used at Freeman for sealing off moisture in many of the chemical batches Freeman manufactures. Airco was called for instructions on how to stop the leak. Instructions were received to turn off a set of valves.

Visibility zero

A three-man crew in full turnout gear, including SCBA, took a 1½-inch line from 365 and entered the vapor cloud and tank area. Prior to this time, no one was permitted near the vapor cloud and everyone was kept upwind. Upon entering the area of the tank, visibility was almost zero, but the men found the control valves and closed them.

Airco asked for a tank reading on the amount of pressure, but this was impossible due to the low visibility. The men then retreated from the area.

Unfortunately, these actions did not stop the leak. Airco was recalled for more instructions. They determined that the tank was overcharged when it was refilled prior to the weekend. Since Freeman had been shut down for the weekend and no liquid nitrogen was being used, the pressure had built up to a point where the safety valve should have functioned.

Later inspection showed that there were two safety valves: a valve which bleeds off periodically when extensive pressure develops and a safety disk which is designed to rupture when the bleed-off valve does not expel enough pressure. However a valve leading to the bleed-off valve was not reopened after refilling the tank. When excessive pressure developed, the safety disk worked properly and ruptured.

Disk worked . . . and ruptured

Our instructions then were to close the valve leading to the safety disk that ruptured. In order to continue relieving the excessive pressure so the tank would not heat up and BLEVE, the liquid nitrogen was routed into the plant to waste, so that the tank's pressure would continue to be relieved after the safety disk valve was closed.

We were two hours into the incident. All master streams were still operating and dispersing the vapor cloud. The liquid nitrogen vapor had built up a large layer of ice and frost on the valves and piping. It was decided to melt the ice formation from the valves and pipes with a 1½-inch line on straight stream. Again, it was imperative that we not put any water on the tank, because that action would actually heat the tank.

Defrosted the valve

The master streams continued to operate, and the defrosting with the 1½-inch line commenced. After a half hour, it appeared the icing had melted. A 24-foot roof ladder was raised to the tank where the valve shutoff was located. Three men in full turnout gear and SCBA ascended the ladder to close the valve. It wouldn't budge.

They descended and the defrosting with the 1½-inch line continued. After 20 minutes, they again ascended the ladder armed with a pipe wrench. This time they were able to free the valve and ultimately stop the leak. The incident was under control at 5:50 a.m. and all lines were shut down. Sixty fire fighters were used to control this incident.

We feel that we learned several important things and the pre-fire plan provided for much of the success we gained from this incident. We also feel that it is necessary to list them.

1. Nearly 500,000 gallons of water was used in this incident. We knew in advance that we had a good water system and that we could easily get all the water we needed. "Know your resources."

2. All fire fighters must use full turnout gear and SCBA near the incident. After an extended period of time, some fire fighters lost their respect for the hazard. Our chief officers kept close tabs and control of the area so that our personnel did not become casualties.

3. Cooperation from and between Freeman's supervisory personnel was excellent. They know what to expect. Let the experts provide the technical advice before acting on a hunch. If they don't have the answer, they usually know where to get one.

4. Have a plan formulated for automatic move-up if all your equipment is committed. If the call for the house fire had been received five minutes later, our third engine would have been committed. Our second-alarm response provides for a standby engine in our station from a neighboring community.

5. The use of the hazardous materials guide enabled us to make an immediate decision and list the priorities we should follow. With this information were able to control the incident. With the technical advice of Freeman and Airco, we were able to eliminate the hazard.

6. Have plenty of filled replacement air tanks available. Know where to get more. Establish a location upwind where they can be gathered and changed, but nearby so you do not go far for changing.

7. Be careful. Don't do something that could ruin your whole day! ☐ ☐